Teubner-Reihe Wirtschaftsinformatik

A. Oberweis/H. M. Sneed (Hrsg.)

Software-Management '97

Teubner-Reihe Wirtschaftsinformatik

Herausgegeben von

Prof. Dr. Dieter Ehrenberg, Leipzig
Prof. Dr. Dietrich Seibt, Köln
Prof. Dr. Wolffried Stucky, Karlsruhe

Die „Teubner-Reihe Wirtschaftsinformatik“ widmet sich den Kernbereichen und den aktuellen Gebieten der Wirtschaftsinformatik.

In der Reihe werden einerseits Lehrbücher für Studierende der Wirtschaftsinformatik und der Betriebswirtschaftslehre mit dem Schwerpunktfach Wirtschaftsinformatik in Grund- und Hauptstudium veröffentlicht. Andererseits werden Forschungs- und Konferenzberichte, herausragende Dissertationen und Habilitationen sowie Erfahrungsberichte und Handlungsempfehlungen für die Unternehmens- und Verwaltungspraxis publiziert.

Software-Management '97

Fachtagung der Gesellschaft für Informatik e.V. (GI), Oktober 1997 in München

Herausgegeben von

Prof. Dr. Andreas Oberweis
Johann Wolfgang Goethe-Universität
Frankfurt am Main

und Harry M. Sneed
SES Ottobrunn

B. G. Teubner Verlagsgesellschaft
Stuttgart · Leipzig 1997

Prof. Dr. Andreas Oberweis

Geboren 1962 in Trier. Von 1980 bis 1984 Studium des Wirtschaftsingenieurwesens an der Universität Karlsruhe. Wissenschaftlicher Mitarbeiter an der Universität Karlsruhe (1985), an der Technischen Hochschule Darmstadt (1986–1987) und an der Universität Mannheim (1987–1990). Juli 1990 Promotion an der Universität Mannheim mit einer Arbeit über Zeitstrukturen in Informationssystemen. Wissenschaftlicher Assistent am Institut für Angewandte Informatik und Formale Beschreibungsverfahren der Universität Karlsruhe (1990–1995). Februar 1995 Habilitation für das Fach Angewandte Informatik ebendort. Seit 1995 Inhaber eines Lehrstuhls für Wirtschaftsinformatik an der Johann Wolfgang Goethe-Universität Frankfurt am Main.

Harry M. Sneed

Geboren 1940 in Mississippi (USA). Studierte an der Universität Maryland (USA), wo er 1969 sein Master Degree in Public Administration erwarb. Von 1969 bis 1971 als Programmierer/Systemspezialist bei der US-Navy. Von 1971 bis 1979 zunächst beim Hochschulinformationssystem in Hannover, dann bei Siemens in München als Systementwickler. 1980 gründete und leitete er die Firma SES Software-Engineering Service in Ottobrunn/München, die er seit 1993 als Technischer Direktor unterstützt. Von 1987 bis 1990 leitete er zudem das Software-Entwicklungslabor in Budapest. Autor von sieben Büchern zu diversen Themen des Software-Engineering und mehr als achtzig wissenschaftlichen Fachbeiträgen zu den Themen Software-Test, -Wartung, -Reengineering, -Metriken und -Management.

Gedruckt auf chlorfrei gebleichtem Papier.

Die Deutsche Bibliothek – CIP-Einheitsaufnahme

Software-Management '97:
Fachtagung der Gesellschaft für Informatik e.V. (GI), Oktober 1997 in München /
hrsg. von Andreas Oberweis und Harry M. Sneed. –
Stuttgart ; Leipzig : Teubner, 1997
(Teubner-Reihe Wirtschaftsinformatik)
ISBN-13: 978-3-8154-2603-6 e-ISBN-13: 978-3-322-85166-6
DOI: 10.1007/978-3-322-85166-6

Umschlaggestaltung: E. Kretschmer, Leipzig

Vorwort

Software-Management umfaßt alle Management-Aktivitäten für die Software-Entwicklung und den Software-Einsatz. Ein effizientes Software-Management mit zuverlässigen Zeit- und Aufwandsprognosen ist wichtige Voraussetzung für die erfolgreiche Entwicklung und den wirtschaftlichen Einsatz von Informations- und Kommunikationstechnologien in Unternehmen. In Zeiten einer zunehmenden Unternehmensvernetzung im Internet stellen sich für das Software-Management ganz neue Herausforderungen.

Der vorliegende Tagungsband enthält eine Auswahl von Beiträgen der Fachtagung Software-Management '97 vom 29. bis 31. Oktober 1997 in München, veranstaltet vom Fachausschuß 5.1 "Management der Anwendungsentwicklung und -wartung" in der Gesellschaft für Informatik e.V. (GI).

Zu diesem Fachausschuß gehören die drei Fachgruppen

- *Vorgehensmodelle für die betriebliche Anwendungsentwicklung,*
- *Projektmanagement,*
- *Reengineering und Wartung betrieblicher Anwendungssysteme.*

Insgesamt wurden für die Tagung 30 Beiträge eingereicht, die jeweils von drei Mitgliedern des Programmkomitees begutachtet worden sind. Von den eingereichten Beiträgen wurden 14 Beiträge für diesen Tagungsband ausgewählt. Die Beiträge können inhaltlich folgenden Teilbereichen des Software-Managements zugeordnet werden:

- *Projektmanagement*
  (Fell),
- *Experience Factory - Strategie und Bewertung*
  (Landes/Schneider, Lukassen, Stickel),
- *Wartung*
  (Dömer, Hildebrand, Wirdemann),
- *Werkzeuge*
  (Henrich, Schollmeyer/Müller-Luschnat),
- *Erfahrungsberichte*
  (Buxmann/König/Rose, Häck),

- *Workflow-Management und Prozeßautomation*
  (Dellen/Holz/Maurer/Pews, Grünbacher/Hofer, Holten/Striemer/Weske).

Zusätzlich zu den genannten referierten Beiträgen enthält dieser Tagungsband drei eingeladene Beiträge (Elzer, Knolmayer/Spahni, Angele/Studer), die sich mit neuen Anforderungen an das Software-Management, mit der "Jahr-2000-Problematik" sowie mit einem Methodenvergleich zwischen Software Engineering, Knowledge Engineering und Informations Systems Engineering befassen.

Wir möchten an dieser Stelle den Mitgliedern im Programmkomitee für ihre Unterstützung bei der Begutachtung der Beiträge danken:

Prof. Dr. G. Chroust (Uni Linz)
K.B. Elbrechter (Commerzbank Frankfurt/M.)
H.-J. Etzel (Eschborn)
Prof. Dr. H. Heilmann (Uni Stuttgart)
Prof. Dr. W. Hesse (Uni Marburg)
Prof. Dr. K. Hildebrand (FH Ludwigshafen)
Dr. H. Hummel (IABG Ottobrunn)
Dr. R. Kneuper (TLC Frankfurt/M.)
Prof. Dr. G. Knolmayer (Uni Bern)
Prof. Dr. W. König (Uni Frankfurt/M.)
A. Lukassen (COLT Telecom Frankfurt/M.)
K. Manny (CDA Datentechnik Backnang)
PD Dr. A. Meier (CSS Versicherung Luzern)
Prof. Dr. H.C. Mayr (Uni Klagenfurt)
Prof. Dr. W. Mellis (Uni Köln)
G. Müller-Luschnat (FAST e.V. München)
Dr. W.D. Nagl (Oracle München)
Prof. Dr. E. Ortner (TH Darmstadt)
Dr. K. Pohl (RWTH Aachen)
Dr. R. Richter (Uni Karlsruhe)
Prof. Dr. G. Saake (Uni Magdeburg)
Prof. Dr. T. Spitta (Uni Bielefeld)
Prof. Dr. W. Stucky (Uni Karlsruhe)
Prof. Dr. R. Studer (Uni Karlsruhe)
H. Thoma (Novartis Services Basel)

Außerdem danken wir Herrn Oliver Foshag (SES Ottobrunn) und Frau Barbara Wix (Oracle Institut München) für die Unterstützung bei der lokalen Organisation der Tagung.

Frankfurt, München,

im Oktober 1997

Andreas Oberweis

Harry Sneed

Inhalt

Determining Work Units in Year 2000 Maintenance Projects

Gerhard F. Knolmayer, Dieter M. Spahni

Abstract

The dominant issue for software management at the end of this century is to make information systems year 2000 compliant. The respective changes of programs and data stores need very high human and system resources. Not all programs and data stores can be converted at once; therefore, work units that consist of those programs and data stores that will be changed simultaneously have to be determined.

In this paper, a hierarchical planning procedure to construct appropriate work units is developed. It combines a cluster identification algorithm, a branch-and-bound technique and a greedy heuristic. The solution obtained also shows whether the planner can expect the project to be finished in time with the resources available. The procedure has been implemented in the prototype FUSE2000 (Fusion & Sequence); the application of this prototype is illustrated by an example.

1 Introduction

The most striking issue for software management at the end of this century is to make information systems (IS) year 2000 compliant. The strategic decision is between replacement of existing legacy systems, e.g. by state-of-the-art enterprise management systems like Baan, Oracle Applications, or SAP R/3, versus maintenance of these systems. The latter approach usually results in the by far largest software maintenance project which ever hit the organization. From a global point of view there is no doubt that solving the year 2000 (Y2K) problem results in the most voluminous software maintenance effort with which the world was ever confronted.

Y2K projects result from attempts to solve the Y2K problem. The reason for this problem is that in many data stores (data base systems and conventional file organizations) the year is represented by only 2 digits YY; these will usually be numerical values but other data types are also prevalent. Implicitly it is assumed that 1900 has to be added to a two-digit, numerical date element. This design decision has several roots:

- Storage space was far more scarce and costly in the past than it is today. Therefore programmers tried to save storage by using only two digits for representing the year. An ex post cost-benefit-analysis of using the YY notation argues that this may have saved e.g. a company with 10 GB of data in 30 years approx. 200 million USD [Kapp96].
- Programmers provided user-friendly two-digit YY inputs and saved space at scarce output media like screens; furthermore, they often did not decouple user interfaces and data storage and stored the YY representation.
- The data independency principle [ScSt83] forced also more recent applications to use already existing data stores with YY representation.
- The length of the life cycle of IS has often been underestimated. Many system developers did not assume that the systems they were designing or at least parts of them would still be in use at the end of the century.
- Standardization bodies failed in defining unambiguous standards [ISO88] and in convincing project managers and system developers to use the results of their work.

All these issues resulted in a chaotic variety of individual implementations of temporal data and its processing, partially due to

- missing or subjectively unknown standards
- lack of temporal data types and
- diverse temporal reference points associated with different software packages and tools.

Practical experiences and empirical research show that time-oriented data are a component of many applications and data stores (Figure 1) [Rubi96]. About 80% of all programs and about 60% of all data stores are assumed to handle or represent time-varying data. It is assumed that about 3% of all code makes references to temporal data and about 25% of the temporal data resp. processing is infected. Thus, almost 1% of the code has to be changed in Y2K projects [Rubi96]. Some portions of this code have not been touched for many years.

Furthermore, some industries like insurance companies use temporal data more intensively than others. It is important to recognize that, even in a given program, years are often represented partially by the YY and partially by the YYYY notation and that in many cases the windowing technique described in Section 2 has already been implemented not for the sake of solving the Y2K problem but according to the underlying application logic.

	Companies		
Properties of Information Systems	**Financial Services Processing Company**	**Insurance Company**	**Manu-facturing Company**
Number of programs	5'285	14'085	14'930
Lines of Code	6'500'000	14'000'000	19'000'000
Data stores	24'034	32'785	7'195
Temporal data elements	331'761	1'138'950	1'072'631
Infected temporal data elements	89'535	243'313	182'842
Calculations with temporal data	58'157	62'586	39'830

Figure 1: Results of case studies on the dimensions of Year 2000 problems

Many problems and errors may result in using IS if the infected YY representations of years are not changed before approaching the year 2000:

- The determination of intervals may lead to negative values if e.g. instead of 97-93=4 the system computes 00-96=-96. Some programmers may have decided that a sign is not necessary for computing interval lengths, thus resulting in |00-96|=96.
- Wrong intervals may lead to many problems e.g. in interval-dependent computation of costs or benefits (e.g., computing the interest for a bank account).
- Cutting the leading two digits in YYYY leads to wrong sorting sequences, e.g. 00 01 97 98 99. Thus, tables or graphics may represent time series data in a misleading sequence and lead to erroneous conclusions.
- The age of data may trigger some operations in the data stores, e.g. trigger delete procedures for data elements that are older than 30 years. Thus, when the most recent data appear as 00, they will be interpreted as 1900 and may start a delete procedure for these data.
- An expiration date 2000 may be interpreted analogously as 1900 and lead to the rejection of very recent products.

Some of these problems will occur before the year 2000 because different applications anticipate different event horizons. Minor problems with representing the year 2000 as 00 already occurred e.g. in the 70s when insurance companies were unable to handle policies with a duration of 30 years.

In the following we assume that programs will work correct if the year is represented by YYYY. Additional problems of the year 2000, resulting e.g. from being a leap year (according to a twofold exception in the rules of the Gregorian calendar) or because the date element is misused as control information, are not considered in this paper.

2 Methods for solving the Year 2000 problem

One may try to avoid the complexity of Y2K projects and switch to state-of-the-art software packages that have been either designed with the Y2K problem in mind or already been made Y2K compliant [Walt97]. The more powerful such a package is, the less difficult are the remaining problems to solve: The kernel of the IS architecture is provided by the enterprise management system and only direct interfaces to this kernel have to be changed and tested.

Another policy may be to neglect the Y2K problem and wait until the problems manifest. This may be appropriate for applications where avoiding the YY representation is only of cosmetic value. However, this policy is extremely risky for mission critical systems because so many errors may occur around the date change that these cannot be corrected in adequate time horizons, result in a severe derogation of business processes, may lead to loss of reputation and negatively influence the competitive position of the enterprise. „Companies which don't fix the problem won't necessarily fail by the turn of the century, but will be less efficient and therefore less profitable" [Essi97]. It has even been forecasted that 30% of all companies will have to go out-of-business due to their inability to solve the Y2K problem in time [FaLa95].

In the following we concentrate on Y2K projects which try to maintain existing legacy systems. Several maintenance methods have been proposed for solving the Y2K problem:

- Expansion of the year format from YY to YYYY in the data stores. This method seems to be straightforward and is the only one which will perform reliable in the long run; however, it is usually regarded as the most resource-intensive method because programs as well as data stores have to be changed.
- Using (application) logic for algorithmic determination of the 4-digit numerical representation. With this method 100 years are split in p past years, 1 present year, and 99-p future years. This technique can either be implemented with fixed or with sliding windows [IBM97]. When applying window techniques one has to select the window parameter p. In particular one must decide whether p can be defined enterprise- or application- or program-wide or whether individual window lengths have to be defined for each date element of a record.
- Transformation of the YY representation. It has been proposed to

- pack the YYYY representation in such a way that it can be stored in a two-digit field
- pack the YY representation together with a flag (e.g., 0 for 1900 and 1 for 2000) to facilitate the computation of the 4 digit year representation
- use the months 13 to 96 together with YY=99 for representing the years 2000 to 2006
- use character representations in YY e.g. for a maximum of 262 years (if a 26 character set is used)
- encapsulate the YY in such a way that for internal storage the year YY is transformed to XX by

$$XX = \begin{cases} YY - C & \text{for } 99 - p - C < YY \leq 99 \\ 100 + YY - C & \text{for } YY \leq 99 - p - C \end{cases}$$

  The constant C should be chosen in such a way that calendar based operations suffer as little as possible by the transformation. Thus, to keep the day-of-the-week- and the leap-year-determination correct, one should use a multiple of 7*4=28 as C. Usually, C=28 is proposed. However, e.g., the temporal position of holidays will not be correct after this transformation.

A comparison of relevant properties of these methods is given in Figure 2 (for other comparisons cf. [DBSS96; ElLo96; IBM97; Turn97]). Some companies resp. consultants have reported first cost comparisons:

Gartner Group: 1.65 USD per executable line of code for field expansion
1.10 USD per executable line of code for windowing [Brow97].

New York City Transit Authority: Building windows saves 75% of costs that occur with date expansion [Cohe96].

Many Y2K managers are afraid that constraints on time, human, and financial resources do not allow to realize the expansion of all infected data elements. Thus, many organizations will probably combine several methods for making different applications and data stores Y2K compliant.

Property	Expansion YY → YYYY	(Sliding) Windows	Transformations Packing	Encapsulation
Change of temporal data stores				
structures	yes	no	no	no
data types	no	no	yes	no
content	yes	no	yes	yes
Treatment of historical data	also expansion or special treatment	no additional problem	also transformation or special treatment	
Change of programs	yes	yes	yes	yes
Test effort	large	very large	very large	very large
Applicability	general	if relevant time interval < 100 years	very unfriendly to humans	unfriendly to humans
Storage volume				
long run	larger	unchanged	unchanged	unchanged
during conversion	larger	unchanged	larger	larger
Performance	slightly worse	worse	in most cases: worse	slightly worse
Maintainability beyond year 2000	good	bad	very bad	bad
Solution horizon	long run	short run	short run	short run

Figure 2: Comparison of main methods for solving the Year 2000 problem

3 Work Units, Bridges and Application Program Interfaces

One of the properties of the Y2K problem is that a very large number of infected applications and data stores is touched by this maintenance effort; on the other hand, the single maintenance operation for a correctly identified date is rather trivial. Due to the huge number of affected programs and data stores it becomes clear that not all of them can be changed in a big bang operation at a certain time-point: "All the changes required to fix the Y2K problem will be very difficult to manage even if spread over the next few years. All at once would be a disaster" [NN97c].

Work units consist of closely related programs and data stores that are put into production simultaneously. A very detailed phase model for Y2K projects proposes a step

"2.1.7.1 Manageable Sub-Projects: Organize conversion into manageable sub-projects and prioritize" [Vald96].

The Y2K project manager of Chubb Insurances stated: "We also looked at major deliverables that were coming down the road in the next 18 months that might cause us to put a particular system either early or late in the renovation schedule. We tried to group similar systems so that we wouldn't have to build lots of bridges when we put the renovated code back into production" [Jung96]. And, more generally, it is recommended: "You manage risk of the conversion by putting in smaller, non-dependent, systems or sub-systems with the benefit of being able to define where anomalies are occurring in your conversions" [Cook96].

Complex IS consist of a large number of interfaces e.g. between

- different modules of a program
- different programs of an application
- modules and data stores
- systems between that data are transferred electronically
    - in-house

- between the boundaries of organizations.

Usually, several or even many programs access a certain data store. Therefore, it will be impossible to construct work units without building interfaces which may materialize e.g. as bridges, application program interfaces (API), redundant data stores or even as redundant programs [UlHa97, p. 128,RICO97]. A bridging program enables e.g. that

- an application program has already been changed to the YYYY representation and some of the data stores used have been converted, too, but it also accesses data stores which still use the YY representation and these temporal representations will be expanded sometime in the future from YY to YYYY
- several application programs access a data store which has already been changed to the YYYY representation and some of the programs already apply the YYYY representation whereas others still use YY and these programs will be changed sometime or other.

The number and the complexity of the bridges to be built transitorily depends on the elements of the work units. "Bridge programs are temporary solutions to daunting problems ... Bridge programs are messy, inelegant, and prone to strange failure modes. But ... they can keep things moving while underlying problems are fixed" [Ashl96, p. 66; cf. also Keog97, p. 169]. Recently, several bridging techniques are implemented in Y2K tools; e.g., VIASOFT offers BRIDGE2000 which allows forward bridging by transforming unconverted data for use in converted programs and reverse bridging for use of converted data stores by non-converted programs and, thus, the coexistence of converted and non-converted system components [VIAS96; VIAS97]. More generally, the importance of "bridgemasters" is stressed who have to understand the bridging products on the market, to test them, to look at all opportunities to do bridges and to document the best practices to do bridging [NN97a].

"Interface issues can be minimized by consolidating closely tied applications into a single project" [Butl96, p. 85]. The importance of building work units was also emphasized in contributions a Y2K-mailing list [e.g. by D. Estes 1996-12-26 ("Only at this point do I start to consider segmenting the programs. Sometimes, they segment easily, with a few bridge points. Other times, there is no optimal segmentation, and in extreme cases the only segmentation is the whole library") or B.L. Johnson 1996-09-25 ("Also think about a tool to help you determine which entities you convert at the same time because the extensive interrelationships between entities")]. Millennium Solutions states: "You will probably want to phase in the installation of your converted programs. This requires that already converted programs reference expanded files, while programs which are not yet converted reference the old version of those same files. To accommodate this, the file conversion programs we generate are also self-bridging. That is, they have the capability of running stand-alone, for one-time file conversion where appropriate, or they can be used as called modules to perform an on-going bridging function, expanding years following reads, and contracting years prior to writes. Both options use the same conversion logic so that a file can be left un-expanded until all processes which use that file are complete" [MiSo97]. Similar arguments can be found in [IBM97].

In Figure 3 we show an information architecture in which data are exchanged between 3 application programs via APIs. We also assume that each of the 3 application programs accesses all 3 data stores. If we presume that the temporal processing in AP1 is changed from YY to YYYY but AP2 and AP3 still process YY date representations, DS1, DS2 and DS3 must be kept redundant. By applying bridging techniques like VIASOFT's BRIDGE2000, *one* call statement from AP1 to BRIDGE2000 will suffice to retrieve the information that e.g. DS1 still employs the YY notation. The accompanying repository provides information about the parameter values employed in the (fixed) window; this allows to transform the YY value in DS1 to the YYYY value in AP1. However, tools like BRIDGE2000 will not be available for all types of programming languages and data organizations, must be accessible online and therefore provide very high reliability, and are not available for APIs. Thus, if e.g. the data flow from AP1 to AP3 is regarded, one must modify the API to transform the YYYY date representation in AP1 to the YY representation in AP2. This effort could be avoided if AP1 and AP3 would be changed in the same work unit.

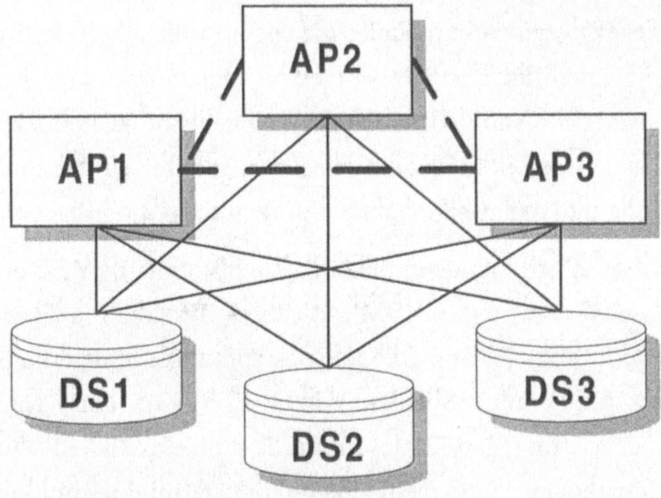

Figure 3: Interfaces between System Components

4 Methods for supporting the determination of work units in Year 2000 projects

4.1 Previous methods

In a posting to a Y2K-forum the problem of constructing work units was characterized as „a scheduling nightmare“ [NN97b]. MS Millennium recommends the following rules for the composition of work units (which they call kernels):

- "Interaction of data between applications components
- Application component inventory & relationships
- ‘Intuitive’ information from application users and subject matter experts
- Production schedules" [MSMi97].

MS Millennium also provides a graphical interface for ad hoc construction of work units in its MSM/2000 tool.

Visionet Systems implemented a proprietary segmentation method in the form of a dependency graph in which programs and data modules constitute the nodes and interdependencies between these modules the edges of the graph. This graph is then re-clustered using a non-linear weighting method. Consequently the system is reduced into a canonical form and partitioned into segments of desired size and properties. Intersections between individual segments are minimized due to the canonical reduction, resulting into a minimum number of required data-bridges [Visi97].

In our approach we relate the determination of work units in Y2K projects to the construction of subsystems e.g. in IBM's Business Systems Planning (BSP) resp. Information System Study (ISS) methodologies and, thus, in determining information architectures (cf. [Brat92; SpHi93; Vett93; KnSp93]). Early contributions to BSP e.g. by James Martin [Mart82] described the problems and advantages of building subsystems without a sound methodology for determining solutions. Formally, the problem can be regarded as a quadratic assignment problem. In BSP, processes and data classes are used as rows resp. columns of the BSP matrix; the non-blank elements of the matrix are C (create, modify) and U (use). By shuffling rows and columns of this matrix, the planner wants to generate a structure which allows to define subsystems in which all C elements and hopefully many U elements are inside the subsystems on the main diagonal whereas U outside these block matrices represent interfaces that should preferably be avoided. Several methods have been proposed to solve this problem:

- ISMOD [Hein85; Katz90]
- A greedy heuristic [Spah97]
- A specialized branch-and-bound algorithm [Spah97]
- Mixed-integer programming models [Knol94]
- Genetic algorithms [KnGe97].

Fig. 4 gives an example of a BSP matrix in original and in rearranged form. The resulting six subsystems and the data flows between them form an IS architecture.

Processes \ Data Classes	Customer	Order	Vendor	Product	routings	Bill of Material	Cost	Parts Master	Raw Material Inventory	Final Goods Inventory	Employee	Sales Territory	Financial	Planning	Work in Progress	Facilities	Open Requirements	Machine Load
Business Planning		U					U						U	C				
Organization Analysis		U												U				
Review and Control		U											U	U				
Financial Planning		U									U		U	C	U			
Capital Acquisition		U											C					
Research				U								U						
Forecasting	U			U								U		U				
Design and Development	U			C		U		C										
Product Specification Maintenance			U	U		C		C										
Purchasing			C				U											
Receiving			U						U									
Inventory Control									C	C					U			
Workflow Layout				U	U											C		
Scheduling			U	U											C	U		U
Capacity Planning			U		U											U	U	C
Material Requirements			U	U		U											C	
Operations					C										U	U	U	U
Territory Management	C	U		U														
Selling	U	U		U								C						
Sales Administration		U										U						
Order Servicing	U	C		U														
Shipping		U		U						U								
General Accounting	U		U								U		U					
Cost Planning		U	U				C											
Budget Accounting							U				U		U	U	U			
Personnel Planning											C		U					
Recruiting/Development											U							
Compensation											U		U					

Processes \ Data Classes	Raw Material Inventory	Final Goods Inventory	routings	Work in Progress	Facilities	Open Requirements	Machine Load	Customer	Order	Product	Bill of Material	Parts Master	Sales Territory	Vendor	Cost	Employee	Financial	Planning
Inventory Control	C	C		U														
Workflow Layout			U		C					U								
Scheduling				C	U		U			U				U				
Capacity Planning			U		U	U	C							U				
Material Requirements						C				U	U			U				
Operations			C	U	U	U	U											
Research										U			U					
Forecasting								U		U			U					U
Design and Development								U		C	U	C						
Product Specification Maintenance										U	C	C		U				
Territory Management								C	U	U								
Selling								U	U	U			C					
Sales Administration									U				U					
Order Servicing								U	C	U								
Shipping		U							U	U								
Purchasing														C	U			
Receiving	U													U				
Cost Planning									U					U	C			
Business Planning									U						U		U	C
Organization Analysis									U									U
Review and Control									U								U	U
Financial Planning				U					U							U	U	C
Capital Acquisition									U								C	
General Accounting								U						U		U	U	
Budget Accounting				U											U	U	U	U
Personnel Planning																C	U	
Recruiting/Development																U		
Compensation																U	U	

Figure 4: Original and rearranged BSP matrix

4.2 Building work units by FUSE2000

4.2.1 Introduction

Application programs have different impact on the day-to-day operations of the organization, for the support of managerial decision making, and, thus, for the competitive position of the organization. With respect to the time criticality by which many Y2K projects are beaten, a triage is necessary to concentrate the effort and resources on those applications which are extremely or very important for contingency of the business processes [deJa96].

Another aspect which influences the sequence in which the programs have to be converted is their expected time to failure. Ceteris paribus, those applications which will fail, e.g., in January 1998 have to be maintained earlier than those in which the year 2000 becomes relevant the first time at the end of 1999.

In the following we assume that a certain milestone finishes the next planning period in a Y2K project. We assume that several time points for conversion of work units are planned in this interval. The number of these time points can be fixed or may be restricted by lower and upper bounds.

We assume that bridging mechanisms have to be realized between programs resp. data stores which are converted at different time points independently from the length of the time interval between the conversion points. Thus, no manual patches are allowed to avoid the construction of bridges.

4.2.2 A repository for temporally infected programs and data stores

A main step in the early phases of a Y2K project is an impact analysis [ArBo93; QVWM94]. The code is scanned for variables and/or the data stores are scanned for elements representing temporal information. The result of an impact analysis are tables in which (hopefully all and only) those programs which use temporal data are associated with temporal data or other programs with which they share temporal data. Such meta-data is often stored in repositories. In the following we assume that a perfect impact analysis has provided all relevant data in a repository in such a way that the interfaces between infected system components can be easily identified. Those programs that use two-digit variables for representing years and/or that access date elements with YY representations are called candidate programs. Data stores that save the year in YY representation are called candidate data stores. First, we construct an adjacency matrix which shows the interaction between candidate programs represented in the rows and candidate data stores represented in the columns of the adjacency matrix; 0 indicates that no interaction exists and 1 symbolizes interaction. APIs that pass infected data can be treated in the same way; for ease of presentation, this type of interfaces will be neglected in the rest of the paper.

4.2.3 Providing data for the determination of work units

4.2.3.1 Complexity measures for interaction between system components with respect to temporal data

In this step of our method we substitute the 1 of the adjacency matrix by a positive value which symbolizes the complexity of the interaction between system components. The complexity is influenced, e.g., by the number of accesses to infected temporal data, the number of infected date and/or code elements, flexibility and maintainability of the infected data stores and the costs of establishing a bridging technique between infected programs and infected data stores.

4.2.3.2 Capacity requirements of transforming the system components

A lot of resources is needed in Y2K projects worldwide and many forecasts predict a considerable shortage of human resources and technical facilities needed for converting and testing of Y2K solutions. Some data stores have to be unloaded, restructured and uploaded to allow the change from YY to YYYY. This can often be done only when the applications are off-line and large databases, e.g. in banks or insurance companies, can often be converted only at weekends. Thus, determining work units must take capacity constraints into account. We allow such constraints for programs as well as for data stores.

4.2.3.3 Applying a cluster identification algorithm for constructing base blocks

In the next step we define base blocks. They contain a certain set of programs and a certain set of data stores which are tightly coupled. A base block is the smallest element that may build a single work unit. However, usually several base blocks are combined to form a work unit.

A Cluster Identification Algorithm (CIA) has been developed for supporting group technology decisions in manufacturing planning [KuCh87]. The CIA allows to determine base blocks, i.e. programs and data stores which are not allowed to be elements of different work units. The number and the sizes of the base blocks are influenced by the determination of a threshold parameter which must be provided by the planner with respect to the complexity matrix. The corresponding element is set to 1 if the matrix element is greater than the threshold; else it is set to 0. The resulting pattern of binary values is used by the CIA to construct the base blocks. The higher the threshold, the more base blocks result. The more base blocks exist, the more flexibility is available in determining the work units. By a modification of the standard procedure it is possible to avoid that heterogeneous components are grouped into the base block; e.g., it may be necessary that all programs in a base block should be written in the same language to allow the application of a certain conversion tool.

4.2.3.4 Determination of the critical base blocks

Priority ranking with respect to business importance and time criticality is used to determine those base blocks which have to be transformed within the next planning period. The set of these critical base blocks is called BBCRIT.

4.2.3.5 Determine the optimal subsystem for the next planning period

All programs and data bases which belong to BBCRIT should be transformed within the planning horizon. However, the capacities available inside the company or externally may allow additional programs and/or data stores to be transformed in the planning period regarded. This step of the procedure is based on the original, temporally-oriented complexity matrix. To make sure that all elements of BBCRIT are assigned to the subsystem SS1 which is converted in the planning period, the complexity coefficients between elements belonging to BBCRIT are set to a very large value M. Those base blocks which are too heterogeneous to be combined obtain a very small value -M.

The assignment procedure has to take into account the capacity constraints resulting from the data determined in Step 6.2.

Those base blocks that are not assigned to SS1 belong to subsystem SS2. They are candidates for conversion in time intervals which lie beyond the horizon of the planning period regarded. Because they will be converted in succeeding periods, the procedure resumes with step 4.2.3.4 for all base blocks within SS2 when determining the work units for the next planning period.

Several mathematical procedures may be considered for the assignment, e.g., a greedy heuristic, optimization procedures based on branch-and-bound or, more generally, on mixed-integer programming, or heuristics based on genetic algorithms. In the prototype FUSE2000 we employ a sophisticated combination between the greedy heuristic and a branch-and-bound approach, similar to the procedure described in [Spah97].

4.2.3.6 Determination of work units for subsystem 1

The assignment procedure described in step 4.2.3.5. is repeated for those base blocks which have been combined to the subsystem SS1. The planner may define the number of work units as the number of conversion steps scheduled for the planning period; e.g., two conversion steps may be assigned until the end of 1997. Alternatively, the planner may also define a lower and an upper bound on the number of conversion steps.

5 An example

Assume that the repository contains 17 candidate programs and 10 candidate data stores (Fig. 5a). The project management team has set up tables with the estimated need of resources for converting these system components (Figure 6). Based on the elements of the complexity matrix with a value greater than the threshold of 5 (shaded gray in Fig. 5a), the CIA determines the 6 base blocks shown in Fig. 5b.

Within the planning period, there are not more than 40 units of resources available for converting programs resp. 30 units for data stores. There are two conversion steps within the planning period, each of them should employ about the same amount of resources.

With respect to business importance and time criticality, we assume that programs P6, P7 and P9 as well as data stores D4 and D7 have to be transformed within the planning period regarded and thus belong to BBCRIT. In order to enforce the fusion of the corresponding base blocks, the complexity coefficients between these elements are set to a very large value M. The resulting two subsystems, each requiring 38 resource units for converting programs and 26 resp. 24 units for enlarging the data stores, are shown in Fig. 7. Subsystem 1 contains all base blocks with M entries and, with respect to the resources available, also program 1 and data store 5. Figure 7 shows that the accesses of P2 and P15 - that will (hopefully) be converted in the next planning period - need bridges to the data store D5 which will be expanded in the first period.

Finally, the two work units within subsystem 1 are determined by the branch-and-bound procedure. They require 18 resp. 20 units for converting infected programs and 12 resp. 14 units for converting infected data stores. After the first conversion step, bridges are necessary for accesses of P1 to D4 and of P7 and P11 to D6 and D7 until the completion of the second conversion step.

	D1	D2	D3	D4	D5	D6	D7	D8	D9	D10
P1				4	8					
P2					1			3	2	7
P3		7	3	3						
P4	6			9						
P5					4	8	7			
P6		6	9	7						
P7				8		1	3			
P8	9			6						
P9					2	6	9			
P10		8	7	8						
P11				6		3	1			
P12								8		
P13										9
P14								3	2	6
P15					3				8	
P16									3	7
P17									8	

Figure 5a: Example of time complexity matrix

	D5	D10	D1	D2	D3	D4	D6	D7	D8	D9
P1	8					4				
P2	1	7							3	2
P13		9								
P14		6							3	2
P16		7								3
P3				7	3	3				
P4			6			9				
P6				6	9	M		M		
P7						M	1	M		
P8			9			6				
P10				8	7	8				
P11						6	3	1		
P5	4						8	7		
P9	2					M	6	M		
P12									8	
P15	3									8
P17										8

Figure 5b: Six base blocks determined by the CIA

P1	P2	P3	P4	P5	P6	P7	P8	P9	P10	P11	P12	P13	P14	P15	P16	P17
8	9	2	4	4	1	3	3	6	6	1	9	1	1	6	7	5

Figure 6a: Resources needed to convert application programs

D1	D2	D3	D4	D5	D6	D7	D8	D9	D10
7	1	2	4	2	7	3	8	7	9

Figure 6b: Resources needed to convert data stores

	D5	D1	D2	D3	D4	D6	D7	D8	D9	D10
P1	8				4					
P3			7	3	3					
P4		6			9					
P6			6	9	M		M			
P7					M	1	M			
P8		9			6					
P10			8	7	8					
P11					6	3	1			
P5	4					8	7			
P9	2				M	6	M			
P2	1							3	2	7
P13										9
P14								3	2	6
P16									3	7
P12								8		
P15	3								8	
P17									8	

Figure 7: Subsystem 1 and Subsystem 2

	D5	D6	D7	D1	D2	D3	D4
P1	8						4
P5	4	8	7				
P9	2	6	9				
P3					7	3	3
P4				6			9
P6					6	9	7
P7		1	3				8
P8				9			6
P10					8	7	8
P11		3	1				6

Figure 8: Work units and bridges within Subsystem 1

6 Summary

In this paper we emphasize the importance of determining work units in Year 2000 projects. This decision highly influences the number of bridges and Year 2000-specific APIs that have to be built temporarily and thus the conversion and testing costs in Year 2000 projects. The prototype FUSE2000 is described in which a cluster identification algorithm, a greedy heuristic and a branch-and-bound algorithm are combined in a hierarchical planning system for construction and sequencing of work units consisting of infected programs and infected data stores. The functionality of FUSE2000 is explained by an example.

References

[ArBo93] Arnold, R.S., Bohner, S.A.: Impact analysis - Towards a framework for comparison. In: D. Card (Ed.), Proc. Conference on Software Maintenance 1993, IEEE Computer Society Press, 1993, pp. 292 - 301.

[Ashl96] Ashley, J.: Year 2000 Solutions. Cambridge Market Intelligence, 1996.

[Brat92] Brathwaite, K.S.: Information Engineering, Vol I, Concepts. CRC 1992.

[Brow97] Brown, D.: Project Management and Project Timelines, EMU & YEAR2000 Seminar, Gartner Group, Zurich 1997.

[Butl96] Butler, J.: The Year 2000 Crisis. Computer Technology Research 1996.

[Cohe96] Cohen, B.: New York Transit Official Poses Y2K Alternatives. In: ITAA's Year 2000 Outlook 1 (1996) 6 (Internet Newsletter) *http://www.itaa.org/scripts/dbml.exe?action=query&Template=2000lett.dbm&IssueID=31 Registration: http://www.itaa.org/scripts/dbml.exe?Template=get2klet.dbm (as of 1997-05-10)*

[Cook96] Cook, W.J.: A Risk Managed Approach to Year 2000 Data Architecture *http://pw1.netcom.com/~wjcook/y2krisk.html (as of 1997-04-23)*

[DBSS96] Dun & Bradstreet Satyam Software (Ed.): Year 2000, Code Enabling Strategies and Problem Sizing. *http://www.dbss.com/y2kpaper.htm (as of 1997-04-23)*

[deJa96] de Jager, P.: Systemic Triage. In: American Programmer 9 (1996) 2, pp. 12-15.

[ElLo96] Eldridge, A., Louton, B.: A Comparison of Procedural and Data Change Options for Century Compliance. *http://www.year2000.com/archive/options.html (as of 1997-05-08)*

[Essi97] Essick, K.: Year 2000 problem may be worse than feared. In: InfoWorld Electric, 1997-03-06. *http://www.infoworld.com/cgi-bin/displayStory.pl?97036.e2000.htm (as of 1997-05-08)*

[FaLa95] Farber, A., LaChance, R.: Impact of the year 2000: Next millennium no chance for celebration. In: Enterprise Systems Journal 10 (1995) 12, p. 36.

[Hein85] Hein, K.P.: Information System Model and Architecture Generator. In: IBM Systems Journal 24 (1985) 3-4, pp. 213 - 235.

[IBM97] IBM (Ed.), The Year 2000 and 2-Digit Dates: Guide Document Number GC28-1251-06, 6th ed., 1997-04-11. *http://ppdbooks.pok.ibm.com:80/cgi-*

bin/bookmgr/bookmgr.cmd/BOOKS/y2kpaper/CCONTENTS(as of 1997-05-08)

[ISO88] ISO 8601: Data elements and interchange formats - Information interchange - Representation of dates and times. Reference number ISO 8601: 1988 (E), (Geneve) 1988.

[Jung96] Jung, J.: Opening statement to CIOs Year 2000 Online Conference, 1996-09-17.
http://www.cio.com/forums/year2k.html (as of 1997-07-01)

[Kapp96] Kappelman, L.: Year 2000 upgrades: A small price to pay. In: Computerworld 30 (1996) 48, p. 33.

[Katz90] Katz, R.L.: Busines/Enterprise Modeling. In: IBM Systems Journal 29 (1990) 4, pp. 509-525.

[Keog97] Keogh, J.: Solving the Year 2000 Problem. Academic Press 1997.

[Knol94] Knolmayer, G.: The Application of Mixed Integer Programming to the „Business Systems Planning"-Problem. In: H. Dyckhoff et al. (Eds.), Operations Research Proceedings 1993. Springer 1994, pp. 457 - 463.

[KnGe97] Knolmayer, G., Gerber, J.-P.: Experiences with applying a genetic algorithm to determine an information systems architecture. In: OR Spektrum 19 (1997) 1, pp. 47 - 53.

[KnSp93] Knolmayer, G., Spahni, D.: Darstellung und Vergleich ausgewählter Methoden zur Bestimmung von IS-Architekturen. In: H. Reichel (Ed.), Informatik, Wirtschaft, Gesellschaft. Springer 1993, pp. 99 - 104.

[KuCh87] Kusiak, A., Chow, W.S.: An Efficient Cluster Identification Algorithm. In: IEEE Transactions on Systems, Man, and Cybernetics 17 (1987) 4, pp. 696-699.

[Mart82] Martin, J.: Strategic Data-Planning Methodologies. Prentice-Hall 1982.

[MiSo97] Millennium Solutions: Legacy System Survival in the Year 2000
http://www.2k-solutions.com/ (as of 1997-05-08)

[MSMi97] MS Millennium (Ed.): The Millennium Challenge & MS Millennium Capabilities, 1997.

[NN97a] NN: Lessons from the trenches: Building - and burning - bridges. In: Computerworld 31 (1997) 18, p. 72.

[NN97b] NN: Reply #364 to VIAS VIASOFT & THE Y2K PROBLEM
http://www.exchange2000.com/~wsapi/investor/s-11784/reply-364 (as of 1997-06-30)

[NN97c] NN: Database-Program inter-dependency.
http://www.bridgeims.com/depend.htm (as of 1997-06-30)

[QVWM94] Queille, J.-P., Voidrot, J.-F., Wilde, N., Munro, M.: The impact analysis task in software maintenance: A model and a case study. In: H.A. Muller, M. Georges (Eds.), Proceedings International Conference on Software Maintenance 1994. IEEE Computer Society Press 1994, pp. 234 - 242.

[RICO97] RICOMM Systems (Ed.): Bridging the Gap, 1997.

[Rubi96] Rubin, H.A.: Millennium Metrics: Truth & Consequences. In: Proc. 10th International Conference on Software Maintenance & Software Management, Year 2000 Solutions, 1996.

[ScSt83] Schlageter, G., Stucky, W.: Datenbanksysteme: Konzepte und Modelle, 2. Auflage. Teubner 1983.

[Spah97] Spahni, D.: Gestaltung von Informationssystemen, Bestimmung und Sequenzierung geeigneter Teilsysteme. Physica 1997.

[SpHi93] Spewak, S.H., Hill, S.C.: Enterprise Architecture Planning, QED 1993.

[Turn97] Turn of the Century (Ed.): Y2K Task Grid. *http://www.tocs.com/y2k/a8.htm (as of 1997-04-23)*

[UlHa97] Ulrich, W.M., Hayes, I.S., The Year 2000 Software Crisis. Yourdon Press 1997.

[Vald96] Valdes, J.: Planning: Y2K Conversion. *Mail to (moderated) list year2000@hookup.net (1996-09-25; sent 1996-10-05)*

[Vett93] Vetter, M.: Strategie der Anwendungssoftware-Entwicklung. Methoden, Techniken, Tools einer ganzheitlichen, objektorientierten Vorgehensweise. Teubner 1993.

[VIAS96] Viasoft: Dynamic Bridging: A Year 2000 Conversion Strategy, BR01BRDG0996, 1996.

[VIAS97] Viasoft: Viasoft's Bridge 2000, BR10BR0497, 1997. *http://www.viasoft.com/PRDCTS/BRIDGE2K/ (as of 1997-05-08)*

[Visi97] Visionet Systems: Millennium/400 Segmentation Tools. *http://www.visionets.com/ (as of 1997-06-30)*

[Walt97] Walter, M.: Vorgehensweise und Erfahrungen bei der Datumsumstellung im SAP System R/2. In: Wirtschaftsinformatik 39 (1997) 1, pp. 19 - 24.

Prioritätensteuerung von Projekten bei der GEZ

Axel Fell

Abstract

In Organisationen, deren Geschäftsprozesse hochgradig mit Informationstechnologie unterstützt werden, existiert zu jedem Zeitpunkt ein bezüglich der organisatorischen Zuordnung und Zielsetzung heterogener Bedarf, Projekte zu initiieren und durchzuführen. Da für die Durchführung nur knappe Ressourcen zur Verfügung stehen, muß eine Priorisierung der Projektwünsche erfolgen werden. Dies erfolgt in der GEZ über ein beschriebenes Beantragungs- und Genehmigungsverfahren mit festgelegten Kompetenzen. Priorisiert wird anhand der Kriterien Wirtschaftlichkeit, Strategische Bedeutung und Operative Dringlichkeit, deren Ausprägung für das jeweilige Projekt in einer dreistelligen Bewertungsziffer abgebildet wird. Das Verfahren ermöglicht es, vergleichend Projektdurchführungen zeitlich und hinsichtlich der Ressourcenzuordnung zu priorisieren, gleichzeitig wird die Dokumentation der für ein Projekt ausschlaggebenden Faktoren vorgenommen. Die Probleme bei der Anwendung des Verfahrens können weitgehend durch Anpassung ohne Strukturbruch beseitigt oder in ihrer Wirkung abgeschwächt werden.

1 Einführung

1.1 Überblick über die GEZ

Die Gebühreneinzugszentrale der öffentlich-rechtlichen Rundfunkanstalten der Bundesrepublik Deutschland ist eine Gemeinschaftseinrichtung der elf Landesrundfunkanstalten und des Zweiten Deutschen Fernsehens zum Zwecke des zentralen Gebühreneinzugs. Sie existiert seit 1.1.1976.

Die Aufgaben der GEZ bestehen insbesondere aus:

- der Pflege der Stammdaten aller Rundfunkteilnehmer

- der Erhebung der Rundfunkgebühren
- der Zahlungsüberwachung
- Arbeiten im Zusammenhang mit Gebührenerstattungen
- der buchmäßigen Erfassung und Abrechnung der Gebührenforderungen, -rückstände und -einnahmen
- der Bestandsführung über Befreiungen von der Rundfunkgebührenpflicht
- der Erstellung von Auswertungen verschiedenster Art über den Gebühreneinzug für die Rundfunkanstalten und die GEZ
- der Durchführung von Maßnahmen zur Hebung des Teilnehmerpotentials

1996 wurden Mrd. DM 9,25 Rundfunkgebühren eingezogen, davon 86% per Lastschrift und der Rest per Einzelüberweisung und Dauerauftrag. Es wurden 36,8 Mio. Teilnehmerkonten geführt, 36,5 Mio. Briefe versandt, die Portokosten alleine beliefen sich auf Mio. DM 16,7. Jeden Monat werden 8.821.000 Lastschriften durchgeführt, ergehen 1.119.000 Zahlungsaufforderungen, ferner 568.000 Erinnerungen, Gebührenbescheide, Mahnungen, Vollstreckungsersuchen und Ordnungswidrigkeitsverfahren. Ebenfalls monatlich werden (ohne Lastschriften) 1.282.000 Zahlungseingänge verbucht, ebenso 76.000 Rücklastschriften und 39.000 Gebührenerstattungen. Jeden Arbeitstag gehen über 52.000 Geschäftsvorgänge Eingangspost bei der GEZ ein.

1.2 Die Organisation der GEZ

Die GEZ besteht aus

- der Geschäftsführung mit den Stabsbereichen
    - Zentrale Aufgaben, Datenschutz
    - Revision
    - Recht
    - Informations- und Öffentlichkeitsarbeit,
- einer Verwaltungsabteilung,
- der Abteilung Verfahrens- und Betriebsorganisation
- dem Geschäftsbereich Gebühreneinzug, Bestandsführung und Abrechnung mit den Fachabteilungen

    - Gebührenabrechnung und Datenbestandspflege
    - Produktions-Controlling, Datenaufbereitung und Datenerfassung
    - Teilnehmerkontenbearbeitung und Telefonberatung
    - Teilnehmerkontenbearbeitung und Mahnmaßnahmen
- dem Geschäftsbereich Systementwicklung und DV-Produktion mit den Datenverarbeitungs- (DV-) Abteilungen
    - Systemprogrammierung und Rechenzentrum
    - Anwendungsentwicklung und -systeme
    - DV-Service und Controlling,

Die GEZ hat ca. 875 Mitarbeiter. Zusätzlich werden im Auftrag externe Dienstleistungsunternehmen in den Bereichen Datenerfassung, schriftliche Sachbearbeitung, Telefonservice und Datenverarbeitung beschäftigt.

1.3 Außenbeziehungen und Kommunikationsverflechtungen der GEZ

Die GEZ unterhält Kommunikationsbeziehungen zu ca. 35,9 Mio. Rundfunkteilnehmern, 12 Rundfunkanstalten, fast 40.000 Geldinstituten (und deren Zweigstellen), 17.000 Postämtern, 15 Postbanken, 1.600 Meldeämtern, ca. 1.400 Sozialbehörden, 735 Vollstreckungsbehörden, 1.947 Ordnungsbehörden und 52 Verwaltungsgerichten sowie weiteren am Rundfunkgebühreneinzug interessierten Institutionen und Personen.

2 Die Problemstellung der Priorisierung von EDV-Projekten

2.1 Die Anwendungslandschaft

Das Gebühreneinzugsverfahren bzw. die Aufgaben der GEZ sind vollständig als (Dienstleistungs-) Produkte und zur Erstellung dieser Produkte erforderliche Geschäftsprozesse definiert. Die Geschäftsprozesse werden hochgradig durch elektronische Datenverarbeitung unterstützt. Die Bedeutung der elektronischen Datenverarbeitung für den Gebühreneinzug spiegelt sich unter anderem darin wieder, daß die DV-Kosten (nach Gemeinkostenumlage) ca. 40% der gesamten Kosten der GEZ ausmachen.

Die Anwendungen für die aufgeführten Aufgaben sind überwiegend Eigenentwicklungen, in einigen Bereichen und für Systemüberwachungs- und -steuerungsaufgaben wird Standardsoftware eingesetzt. Auftragsentwicklungen und schlüsselfertige Systemlösungen sind über zum Teil relativ aufwendige Schnittstellen in die Anwendungslandschaft der GEZ integriert.

2.2 Die personellen Ressourcen

Die zum

- Betrieb
- der Wartung und Pflege
- der Fort- bzw. Neuentwicklung

von Systemen und Verfahren zur Verfügung stehenden personellen Ressourcen sind in den drei DV-Abteilungen und der Abteilung Verfahrens- und Betriebsorganisation zusammengefaßt. Hier sind insgesamt ca. 165 Personen tätig, zusätzlich kann auf ca. 30 Personenjahre externe Unterstützung zurückgegriffen werden. Die externe Unterstützung ist über individuelle Auftragsvergabe weitgehend flexibel steuerbar.

Aus den Fachabteilungen steht Personal zur Aufstellung und Verifizierung von Anforderungen und zur Qualitätskontrolle und Abnahme zur Verfügung.

2.3 Projekte fallen nicht vom Himmel

In der betrieblichen Praxis sind die DV-Abteilungen und die Abteilung Verfahrens- und Betriebsorganisation mit einer großen Anzahl von Wünschen und Anforderungen der Anwender bzw. Nutzer oder sonstiger mit den Gebühreneinzugsverfahren befaßter Personen/Gruppen konfrontiert: neben der Anforderung an einen möglichst reibungslosen Betrieb der Systeme mit hoher Verfügbarkeit und schnellen Antwortzeiten werden Wünsche insbesondere in Bezug auf Änderungen an bestehenden Anwendungen, Verfahren und Systemen oder an deren Neuentwicklung artikuliert. Diese Anforderungen können für die GEZ wie folgt charakterisiert werden:

- Die Anwender- (sprich: Kunden-) Struktur ist heterogen: als Anwender können Externe, Angehörige von externen und internen Gremien oder die Gremien als Ganzes (per gemeinschaftlichem Beschluß), die Fachabteilungen, Angehörige der DV-Abteilungen und der sonstigen Bereiche der GEZ einschließlich der Geschäftsführung auftreten.
- Die Zielsetzungen der Anwender bzw. Anwendergruppen ist ebenso heterogen: von der Vorbereitung bzw. Umsetzungen gesetzlicher, gebührenrechtlicher oder sonstiger verwaltungstechnischer externer Rahmenbedingungen, der Umsetzung von Vorgaben von Fach- oder Verwaltungsgremien, über die Nutzung neuer Technologien bis zur effektiveren und effizienteren Gestaltung der Geschäftsprozesse reichen die potentiellen Auslöser für Anforderungen an die Verfahren bzw. Systeme.
- Da die Zielsetzungen meist aus den Verantwortungsbereichen der Anforderer abgeleitet sind, ergeben sich zwischen den einzelnen Anforderungen nur teilweise komplementäre Zielsetzungen, es treten auch konkurrierende Zielbeziehungen auf.
- Die Art der Äußerung der Anforderungen ist vielfältig: die Veröffentlichung von Gesetzen, die veröffentlichten Beschlüsse der externen und internen Gremien, die formalisierte "Änderungsmeldung" im vorgegebenen Anwendungsentwicklungsverfahren, Ideen von Einzelpersonen oder Gruppen oder auch der Anruf beim Benutzerservice können als Form der Artikulation der Nutzerwünsche auftreten.
- Im Moment der Äußerung der Kundenwünsche ist in der Regel das Ausmaß der zu ihrer Befriedigung erforderlichen Ressourcen nur bedingt abschätzbar, insbe-

sondere für solche Anforderungen, die über einfache und schnelle Änderungen an bestehenden Verfahren hinausgehen.

- Bei komplexeren Anforderungen ist die technische Realisierbarkeit sowie die dabei vorhandenen Optionen zunächst unbekannt.

2.4 Der Bedarf an Prioritätensteuerung

Aus der Heterogenität der Anforderungen sowie dem hohen Maß an Unsicherheit über die Ressourcenbindung und die technische Realisierbarkeit auf der einen Seite, der Beschränkung der Ressourcenverfügbarkeit auf der anderen Seite entsteht der Bedarf, die zur Befriedigung der Anforderungen anstehenden Aktivitäten (Projekte) in Form eines Verfahrens mit definierten Ergebnissen zu priorisieren. Ziele eines solchen Priorisierungsverfahrens sind:

- eine möglichst vollständige Erfassung der Anforderungen und Wünsche, die mit einer gewissen Wahrscheinlichkeit zu Projekten führen,
- größtmögliche Transparenz über die der Implementierung von Projekten zugrunde liegenden Daten bei den Entscheidungsträgern,
- die systematische und formalisierte Dokumentation über die grundlegenden Bestandteile der Anforderungen (Ziele, Bedarfsträger, betroffene Geschäftsprozesse und Produkte, Ressourcenzuordnung, Termine, Verantwortungsbereiche etc.),
- die Definition eines verbindlich einzuhaltenden Entscheidungsprozesses zur Implementierung von Projekten mit klarer Benennung der Instanzen
- die Erarbeitung und Dokumentation von Prioritäten zwischen unterschiedlichen Vorhaben/Projekten um
    - eine grundlegende Entscheidung über Realisierung oder Nicht-Realisierung von Projekten treffen zu können,
    - eine zeitliche Abfolge der Projekte festlegen zu können,
    - eine Zuordnung der zur Verfügung stehenden Ressourcen nach Anzahl und Qualität durchführen zu können,

- die standardisierte Abstrahierung vom eigentlichen Projektziel bei der Priorisierung und somit eine Vergleichbarkeit von Anforderungen von verschiedenen Anforderern mit gegebenenfalls unterschiedlichen Zielsetzungen.

3 Das Priorisierungsverfahren bei der GEZ

3.1 Beantragung von Vorhaben und Projekten

Die Prioritätensteuerung für Projekte bei der GEZ erfolgt über ein Antrags- und Genehmigungsverfahren mit festgelegten Instanzen. Gegenstand der Beantragung und der Genehmigung ist hauptsächlich die Durchführung von Aktivitäten und damit die Bindung entsprechender Ressourcen, daneben jedoch auch die einzelnen Angaben im Antrag selber, wie sie im folgenden dargestellt sind.

3.1.1 Beantragung eines Vorhabens

Die Beantragung eines Vorhabens ist stets der erste Schritt, sich überhaupt mit einer bestimmten Thematik, die einmal in ein Projekt münden könnte, auseinanderzusetzen. Grundsätzlich kann jeder Mitarbeiter der GEZ ein Vorhaben beantragen, der Antrag muß jedoch vom entsprechenden Abteilungsleiter unterschrieben sein. Das bedeutet, daß die Wünsche Externer zunächst von einem GEZ-Mitarbeiter aufgenommen und verarbeitet werden müssen. Die Entgegennahme der Anträge erfolgt zentral durch den Sachbereich DV-Planung und Controlling, der die Vorhaben und Projekte administriert. Der Antrag muß folgende Elemente enthalten:

- Kurzbezeichnung (6 Stellen, entsprechend den Projektkonventionen)
- Langtext (möglichst in einer Zeile zu definieren)
- Kurzbeschreibung des Vorhabens
    - Zielsetzung
    - Vorläufige Angabe der Anwender/Nutzer/Kunden (Kostenstellen)
    - Vorläufige Angabe der Geschäftsprozesse (oder Arbeitsabläufe) und/oder der Produkte (oder Arbeitsergebnisse) der Anwender, die beeinflußt werden sollen

  - Kurzbeschreibung des erwarteten Nutzens (möglichst anhand der erwarteten Auswirkungen auf Geschäftsprozesse, Arbeitsabläufe, Produkte und Arbeitsergebnisse)
  - Kurzbeschreibung der Nachteile bei Nichtrealisierung
  - Alternative Lösungsansätze (falls möglich)
- Aufwand für eine Voruntersuchung (geschätzt)
- Aufwand für eine mögliche Realisierung (geschätzt)
- Beabsichtigter Zeitraum (Beginn- und Endedatum der Voruntersuchung)
- Vorschlag für personelle Zuordnungen (Leiter Voruntersuchung, ggf. mit Angaben über externe Beratungsunternehmen).

DV-Planung und Controlling prüft die Angaben formal und leitet den Antrag schnellstmöglich an ein Entscheidungsgremium weiter, das aus den Abteilungsleitern der GEZ und der Geschäftsführung besteht. Nach der Entscheidung, eine entsprechende Voruntersuchung einzuleiten, wird das Vorhaben in den Projektkatalog, in dem alle aktiven Vorhaben und Projekte verzeichnet sind, aufgenommen. Außerdem werden die entsprechenden Einrichtungen im zentralen Dokumentationssystem der GEZ vorgenommen.

3.1.2 Voruntersuchung und Abschlußbericht

Für jedes Vorhaben ist eine Voruntersuchung durchzuführen. Ohne Vorlage der im Rahmen einer Voruntersuchung erwarteten Ergebnisse kann ein Vorhaben zur Entscheidung für die Realisierung als Projekt nicht zugelassen werden.

Der Abschlußbericht einer Voruntersuchung muß folgende Ergebnisse beinhalten:

- Kurzbezeichnung (6 Stellen, entsprechend den Projektkonventionen)
- Langtext (möglichst in einer Zeile zu definieren)
- Beschreibung der Voruntersuchungsergebnisse
  - Ziel des Vorhabens
  - Anwender des zu entwickelnden Verfahrens/Systems (Kostenstellen)
  - Bei mehr als einem Anwender: "Nutzungsgrad" pro Anwender (als Basis für Kostenverteilungsschlüssel)

  - Geschäftsprozesse (bzw. Arbeitsabläufe) der Anwender, die durch das zu entwickelnde Verfahren/System beeinflußt werden
  - Produkte (bzw. Arbeitsergebnisse) der Anwender, die durch das zu entwickelnde Verfahren/System beeinflußt werden
- Nutzen bei Realisierung (als erwartete Auswirkung des zu entwickelnden Verfahrens/Systems auf Geschäftsprozesse, Arbeitsabläufe, Produkte und Arbeitsergebnisse der Anwender)
- Nachteile bei Nichtrealisierung
- Lösungsalternativen
- Aufwandsschätzungen
- Darstellung der ermittelten Bewertungsziffer (BWZ)
- Vorschlag für eine Projektterminierung
- Vorschlag für die Ressourcenzuordnung
- Unterschrift des Leiters der Voruntersuchung.

3.1.3 Ermittlung der Bewertungsziffer (BWZ)

Die Bewertungsziffer wird aus den Kriterien Wirtschaftlichkeit , Strategische Bedeutung und Operative Dringlichkeit ermittelt [Nag93]. Dadurch wird versucht, allen organisatorischen Anforderungen und Zielsetzungen gerecht zu werden und auch für den Fall, daß nicht wirtschaftliche Projekte dennoch durchgeführt werden, eine durchgängige Dokumentation der Entscheidungsgründe vorlegen zu können.

Zu jedem dieser Kriterien wird eine Punktzahl wie folgt ermittelt:

3.1.3.1 Wirtschaftlichkeit

Die Punktzahl der Wirtschaftlichkeit ergibt sich aus einem gewichteten Mittelwert zwischen je einer Punktzahl für einen quantifizierbaren Nutzen und für einen qualitativen Nutzen-Aspekt. Die Punktzahl für den quantitativen Nutzen wird über eine Amortisationsrechnung der Entwicklungskosten des Verfahrens/Systems über die Differenz der Aufwendungen für das alte und der Aufwendungen für das neue Verfahren/System ermittelt.

Die Punktzahl für den quantitativen Nutzen wird ermittelt aus einer gleichgewichteten Bewertung der Einzelkriterien

- Verbesserung der Aktualität der Informationen
- Verbesserung der Qualität der Informationen
- Erhöhte Sicherheit des neuen Verfahrens
- Reduktion der Komplexität in den Verfahren.

3.1.3.2 Strategische Bedeutung

Die Punktzahl für das Kriterium der Strategischen Bedeutung wird ermittelt durch eine gewichtete Bewertung der Einzelkriterien

- Festigung der Kostenführerschaft
- Kompetenz der GEZ/Ausweitung der Geschäftsfelder
- Sicherung und Steigerung des Teilnehmerpotentials (K.O.-Kriterium)
- Verbesserung des Servicegrads für die Teilnehmer
- Besondere Bedeutung für die Landesrundfunkanstalten bzw. das ZDF
- Verbesserung des Öffentlichkeitsbildes (Ordnungsmäßigkeit, Sparsamkeit, Sozialverträglichkeit)
- Anzahl der vermeidbaren Geschäftsvorgänge reduzieren
- Nutzung neuer bzw. verbesserter Informations- und Kommunikationswege zwischen GEZ, Teilnehmern, Behörden etc.
- Nutzung technischer Entwicklungen zur Zukunftssicherung (K.O.-Kriterium)
- Verbesserung der Arbeitszufriedenheit und Arbeitsleistung.

Hierbei wird jedes Kriterium auf einer Skala von 0 (nicht wahrnehmbare Auswirkungen) bis 5 (signifikante und nachhaltige Auswirkungen/ Verbesserungen) bewertet. Anschließend wird die Summe aus dem Produkt der Bewertungsziffern mit den Gewichtungsfaktoren über alle Einzelkriterien ermittelt, die dividiert durch die Summe der Gewichtungsfaktoren die Punktzahl des Kriteriums Strategische Bedeutung ergibt.

3.1.3.3 Operative Dringlichkeit

Hier ist die Vorgehensweise gleich mit der bei Strategische Bedeutung, jedoch lauten die Einzelkriterien:

- Externe Rechtsvorschriften (K.O.-Kriterium)
- Interne Vorschriften/Anweisungen
- Abhängigkeit zu anderen Projekten
- Probleme im Tagesgeschäft (K.O.-Kriterium)
- Erforderliches Redesign (K.O.-Kriterium)
- Interne organisatorische Veränderungen

3.1.3.4 Die Bewertungsziffer

Bei jedem der aufgeführten Kriterien ergibt sich ein Punktzahl von 1 bis 4. Die Bewertungsziffer ist eine dreistellige Ziffer, in der die Einzelpunktzahlen der Kriterien nebeneinandergestellt werden [Nag93]. Um die Priorisierung leicht erkennen und eine Reihenfolge herstellen zu können, steht die höchste Punktzahl an erster Stelle, die zweithöchste an zweiter Stelle und die niedrigste an dritter Stelle.

Die Kriterien Strategische Bedeutung und Operative Dringlichkeit werden insgesamt mit 4 (also mit der Höchstpunktzahl) bewertet, wenn jeweils mindestens ein K.O.-Kriterium mit 4 bewertet wird.

3.1.4 Begutachtung

Nach der Erarbeitung der Ergebnisse einer Voruntersuchung gemäß Abschlußbericht und Bewertungskennziffer prüft ein Gremium, die sogenannte Arbeitsgruppe Controlling, in der die Bereichscontroller der GEZ vertreten sind, im Rahmen einer Begutachtung die vorgelegten Ergebnisse. Wenn alle Unterlagen vorliegen und gegebenenfalls nach einer weiteren Erörterung, leitet DV-Planung und Controlling die Ergebnisse an die nächste Abteilungsleiter-Sitzung mit einer Empfehlung in bezug auf die Realisierung und die mögliche Ressourcenzuteilung weiter.

3.1.5 Entscheidung (Start Projekt)

Die abschließende Entscheidung wird in einem Gremium, das aus der Geschäftsführung und den Abteilungsleitern der GEZ besteht, getroffen. Folgende Entscheidungen können getroffen werden:

- Entscheidung, das Vorhaben in ein Projekt mit entsprechender Ressourcenausstattung zu überführen und im Zeitraum xx.xx - yy.yy durchzuführen,
- Entscheidung, das entsprechende Projekt auf den Zeitpunkt zz.zz zu verschieben (ggf. mit reduzierten bzw. verstärkten Ressourcen),
- Entscheidung, das Vorhaben mit dem Ergebnis der Voruntersuchung zu beenden.

Hierbei gilt folgende Klassifizierung:

Muß-Projekte

Projekte, die in einem Kriterium den Maximalwert = 4 haben, oder Projekte, die in der Quersumme der Bewertungsziffer mindestens den Wert = 8 erreichen.

Soll-Projekte

Projekte, die in der Quersumme der Bewertungsziffer einen Wert 6 oder 7 erreichen.

Nicht zu realisierende Projekte

Projekte, die in der Quersumme der Bewertungsziffer einen Wert bis max. 5 erreichen.

Fällt die Entscheidung für die Realisierung des Vorhabens im Rahmen eines Projekts, wird das Projekt in die entsprechenden Dokumentations- und Planungssysteme aufgenommen.

3.1.6 Freigabe der Ressourcen

Werden aus Vorhaben (nach abgeschlossener Voruntersuchung und entsprechender Entscheidung) Projekte, müssen diese mit entsprechender Ressourcenzuordnung versehen werden. Dabei werden die bisherige Projektplanung und die Empfehlungen aus der AG Controlling berücksichtigt. Möglicherweise müssen andere Projekte oder Vorhaben verschoben werden. Eventuell können auch Ressourcen aus dem Bereich für Pflege und Wartung der bestehenden Verfahren zur Verfügung gestellt werden.

DV-Planung und Controlling macht im Zuge der Begutachtung des Vorhabens gegebenenfalls für die bevorstehende Entscheidung Vorschläge, wie die künftige Ressourcenverteilung modifiziert werden kann, um das anstehende Vorhaben als Projekt realisieren zu können.

Durch die Ressourcenzuordnung für dieses neue Projekt können sich bei anderen Vorhaben und Projekten entsprechende Änderungen der Ressourcenverteilung (Kürzungen oder Verschiebungen) ergeben. Diese Maßnahmen werden von DV-Planung und Controlling laufend mit den betroffenen Projektleitern abgestimmt und in einer Gesamtplanung festgehalten.

4 Erfahrungen, Problemstellungen und Optimierungsmöglichkeiten

4.1 Organisatorische Maßnahmen

Das Verfahren zur Beantragung und Bewertung von Vorhaben und Projekten wurde Mitte 1995 implementiert. Es wurde schriftlich gefaßt und von der Geschäftsführung genehmigt. Eine formelle Fassung als Arbeitsanweisung wird derzeit vorbereitet. Zum Zeitpunkt der Einführung wurden alle aktiven Projekte bewertet und in eine Rangfolge gebracht. Seit diesem Zeitpunkt durchlaufen alle potentiellen Projekte das beschriebene Verfahren, es wird weithin bei den Verantwortlichen und Entscheidungsträgern akzeptiert.

Das Verfahren wird vom Sachbereich DV-Planung und Controlling inhaltlich und konzeptionell betreut und administriert.

4.2 Zielerreichung

Das Ziel, mehrere Projekte bereits in der Initiierungsphase hinsichtlich der zeitlichen Abfolge und der Ressourcenzuteilung zu priorisieren, kann als erreicht angesehen werden.

Auch dem Dokumentationsziel wird Rechnung getragen, insbesondere, seitdem in der jüngsten Fassung der Bezug von Vorhaben und Projekten zu Geschäftsprozessen und Produkten der GEZ als Pflichtbestandteil des Projektantrags gefordert wird und für die Wirtschaftlichkeitsrechnungen herangezogen werden soll.

Insgesamt hat sich das Bewertungsverfahren bewährt, der Umgang mit Projekten in der GEZ und die Zuteilung von Ressourcen ist transparenter, effizienter und effektiver geworden.

4.3 Probleme

4.3.1 Die Abgrenzungsproblematik

Ein Problem besteht darin, daß unter dem Verfahren nur diejenigen Aktivitäten erfaßt werden, die aufgrund der Komplexität, der Dauer und der Ressourcenallokation die Voraussetzungen für ein Projekt erfüllen. Nicht erfaßt werden Anforderungen, die im Rahmen von "normalen" Wartungs- und Pflegetätigkeiten an bestehenden Verfahren/Systemen abgearbeitet werden. Hier besteht eine gewisse Gefahr, daß Ressourcenallokation auch in außergewöhnlichem Umfang ohne Vergabe einer Priorität erfolgen kann. Andererseits ist eine Unterwerfung sämtlicher Wartungs- und Pflegearbeiten unter das Bewertungsverfahren viel zu aufwendig. Dieses Problem ist in der GEZ noch nicht abschließend gelöst.

4.3.2 Die Dominanz der Kriterien "Strategische Bedeutung" und "Operative Dringlichkeit"

Insbesondere bei der Bewertung von Vorhaben und Projekten haben sich auch Schwachstellen gezeigt. Es stellte sich heraus, daß es einfacher ist, Projekte über die K.O.-Kriterien unter „Strategische Bedeutung" und "Operative Dringlichkeit" in die Klasse "Muß-Projekte" einzugruppieren, als über den beschwerlichen Weg

des Nachweises der Wirtschaftlichkeit. Dies führte dazu, daß ein Großteil der Projekte tatsächlich über diese Kriterien implementiert wurde. Aus diesem Grunde ist das Verfahren zwischenzeitlich so abgeändert worden, daß der Charakter von K.O.-

Kriterien eliminiert wurde und nur noch die rein rechnerische Bewertung zur Entscheidungsfindung herangezogen wird. Daneben ist die Wirtschaftlichkeitsrechnung zum Pflichtergebnis des Bewertungsverfahrens geworden.

Hierbei werden von Seite DV-Controlling einfache, transparente und DV- unterstützte Verfahren zur Wirtschaftlichkeitsberechnung zur Verfügung gestellt und bei Bedarf Beratung und Hilfestellung erbracht.

Es wird darüber hinaus geprüft, durch die Geschäftsführung eine Zielsetzung etwa derart zu implementieren, daß ein gewisser Prozentsatz von Projekten oder der Gesamt- Ressourcenbindung auf Projekte entfallen muß, die unter dem Kriterium Wirtschaftlichkeit mindestens mit 3 bewertet sind.

4.3.3 Die Darstellung der Bewertungszahl

Was die Darstellung der Bewertungszahl betrifft, zeigt sie zwar in der derzeit verwendeten Form schnell und unmißverständlich die Wertigkeit eines Projekts an und erlaubt auch einen direkten Vergleich zwischen verschiedenen Projekten, es geht jedoch die Information verloren, in welchen Entscheidungskriterien die jeweilige Bewertung erzielt wurde. Hierdurch können beispielsweise Auswertungen über "Arten" von Projekten nicht oder nur nach Rückgriff auf die Dokumentation durchgeführt werden. Durch eine Änderung der Form würden jedoch die oben aufgeführten Vorteile verloren gehen, so daß hier noch nach der optimalen Form der Darstellung gesucht werden muß.

Literatur

[Nag93] Kurt Nagel: Praktische Unternehmensführung: Analysen - Instrumente - Methoden, Landsberg/Lech: Verlag Moderne Industrie, Loseblatt-Ausgabe in drei Bänden, 2. Auflage, 1993

SW-Management im Wandel

Peter F. Elzer

Abstract

Die Entwicklung von Software hat sich in den vergangenen 40 Jahren von einer wissenschaftlichen Herausforderung über eine mehr "handwerkliche" Phase zu einer Großindustrie entwickelt. Die dazu nötigen technischen Hilfsmittel und Managementverfahren haben sich - wenn vielleicht auch nicht immer im vorher erhofften Maße - mitentwickelt. Durch den derzeit stattfindenden Übergang in eine neue Phase - die der Verwendung von Standardkomponenten - müssen sie aber überdacht werden. In diesem Beitrag wird versucht, auf der Grundlage der beruflichen Erfahrungen des Verfassers in Industrie und Hochschule einige der schon eingetretenen oder zu erwartenden Veränderungen darzustellen und ihre möglichen Auswirkungen auf das Management von Softwareprojekten abzuschätzen.

1 Einleitung

Wenn man, wie der Verfasser, 30 Jahre auf dem Softwaresektor tätig ist, hat man so einiges miterlebt - auch die Dauer"krise" der Softwareentwicklung, die nie eine "Krise" war, sondern die ganz normale Formierungsphase einer völlig neuen Technologie mit der zugehörigen Industrie. Dieser Prozeß ist vermutlich noch lange nicht abgeschlossen und momentan sind wieder Änderungsprozesse im Gange. Ohne den Vergleich überstrapazieren zu wollen, könnte man sagen, daß die 70-er Jahre in der DV-Branche durch den Übergang vom Handwerk zur Großindustrie gekennzeichnet waren. Von der Mitte der 80-er Jahre bis heute erleben wir daneben die Entstehung einer Industrie, die auf Entwicklung und Verwendung von Standardteilen beruht.

Die Struktur dieser neuen Form der DV-Industrie unterscheidet sich jedoch nach den Erfahrungen des Verfassers von der im Maschinenbau und ähnlichen Disziplinen. Dort werden üblicherweise die Standardteile von mittelständischen Firmen hergestellt, durch verbindliche und weit verbreitete Normen (die üblicherweise von neutralen Organisationen erarbeitet werden) reguliert, und von anderen mittelständischen oder Großfirmen zu Endprodukten zusammengesetzt.

In der DV-Industrie sieht es dagegen momentan so aus, daß die "Standardteile" von größeren oder großen Firmen (teilweise mit Monopolcharakter) hergestellt und von mittelständischen bis kleinen Firmen zu Endprodukten (="Systemen") zusammengesetzt und an Kundenanforderungen angepaßt werden. Ein Grund dafür mag sein, daß wegen der hohen Entwicklungskosten von Software die Hersteller von "Komponenten" (meist gleichzusetzen mit "Standardpaketen") ihre Märkte so anwendungsübergreifend suchen müssen, daß diese sich mit denen der anwenderorientierten Firmen nur wenig überlappen können. Deshalb können oder müssen sie zum Teil überhaupt keine Rücksicht auf ihre einzelnen Kunden nehmen. Standards werden meist von den Komponentenentwicklern oder kleinen Gruppierungen gesetzt. Man könnte sagen, daß deshalb die Anwendungsentwickler extrem fremdgesteuert sind.

Das wirft neue Probleme für das Management von Anwendungsentwicklungen auf, die deshalb in Zukunft bei entsprechenden Überlegungen mit berücksichtigt werden müssen.

2 Bisherige Entwicklung

Zu Beginn - d.h. bis Ende der 60-er und Anfang der 70-er Jahre, die der Verfasser in der Entwicklung von Echtzeitsystemen, Compilern und Betriebssystemen miterlebte - wurde bei der Entwicklung von Software praktisch immer Neuland betreten, gleichgültig, ob es die Anwendungen oder die verwendeten Programmiertechniken betraf. Selbst die Rechnerhardware änderte sich sowohl von Hersteller zu Hersteller als auch von Modelljahr zu Modelljahr mehr als dies heute der Fall ist. Arbeitstechniken und -hilfsmittel, die heute Allgemeinwissen darstellen, mußte sich jedes Entwicklungsteam selbst erarbeiten. Diese Situation wurde von Menschen, die gerne technisches Neuland betreten, als außerordentlich befriedigend empfunden [Elz90]. Das führte zu einer hohen Motivation und damit zu hervorragenden Ergebnissen. Außerdem waren in dieser Anfangszeit die auftretenden Probleme praktisch ausschließlich technischer Natur und konnten durch jeweils neu geschaffene technische Mittel rasch und wirksam gelöst werden.

Daraus ergab sich wiederum, daß man sich sehr schnell an immer größere und komplexere Aufgaben heranwagte. Dabei wurden aber, wie man heute weiß, unbekannte kritische Systemgrößen überschritten. Das führte dazu, daß die Mehrzahl der "Projekte der zweiten Generation" katastrophal endete. Auch dies hat der Verfasser bei einem großen Leittechnikprojekt in der Industrie miterlebt. Es kam der Eindruck auf, in einer "Softwarekrise" zu stecken. Da aber bis dahin technische Hilfsmittel sehr erfolgreich gewesen waren, glaubte man, durch noch ausgefeiltere technische Methoden, vorzugsweise auf mathematischer Basis, die Probleme wieder in den Griff zu bekommen. Dafür wurde der Begriff des "Software Engineering" geprägt [BuR70].

Die im Rahmen dieses Ansatzes seitdem entwickelten Methoden und Werkzeuge haben sich aber - nach nunmehr über 20 Jahren - immer noch nicht wirklich durchgesetzt. Vor allem ist ihr finanzieller Nutzen oft nur indirekt nachweisbar [Küc91]. Ein Hauptproblem, das allen bisher bekannten Methoden des "Software Engineering" anzuhaften scheint, ist ihr "normativer Charakter". Auf Grund von - meist abstrakten - Überlegungen wird ein "Modell" erdacht, dem der Softwareentwicklungsprozeß gehorchen müsse, um erfolgreich zu sein. Zur Realisierung dieses Modells wird eine "Methode" entwickelt, die dann durch "Werkzeuge" unterstützt wird [Rem87]. Ein Hauptproblem dieser - international üblichen - Vorgehensweise ist, daß zwischen den Entwicklern von "Softwareentwicklungsmethoden" - die normalerweise einen akademischen Hintergrund haben - und der Zielgruppe der industriellen Anwender dieser Werkzeuge eine "Kommunikationslücke" besteht, die in den vergangenen 20 Jahren sogar größer geworden ist [Gla97]. Sie rührt hauptsächlich daher, daß es auf Grund einer "Hochschulpolitik", die man nur als verfehlt bezeichnen kann, nicht möglich ist, an Universitäten Softwareprojekte in Angriff zu nehmen, die auch nur entfernt an die Größe und Komplexität heute üblicher Industrieprojekte heranreichen. Dadurch ist es aber nicht möglich, wissenschaftliches

Verständnis für die Probleme aufzubauen, die sich aus der - notwendigen - Komplexität realer Systeme ergeben. Das war in den 60-er Jahren noch nicht der Fall.

Interessanterweise scheint die Situation auf dem Gebiet neuer "Managementmethoden" ähnlich zu sein.

Um einen Beitrag zur Schließung dieser Kommunikationslücke zu leisten und den Erfahrungsaustausch zwischen Entwicklern von "Softwarewerkzeugen" und "Managementmethoden" und ihren Anwendern zu erleichtern, initiierte der Verfasser seit 1986 im Rahmen der IFAC (= International Federation of Automatic Control) eine Reihe internationaler Workshops: die "IFAC/IFIP Workshops on Experience with the Management of Software Projects" [Elz87, MiE88, MoE89, ElH92, ElR 96]. Obwohl die IFIP (= International Federation of Information Processing) diese regelmäßig als "Co-Sponsor" mitträgt, ist ihr Themengebiet bisher hauptsächlich auf die Entwicklung von Echtzeitsystemen in der Automatisierungstechnik beschränkt. Da es sich dabei aber immer um Entwicklungen hohen Schwierigkeitsgrades handelt, scheinen die jeweils gegebenen Zustandsbeschreibungen auch für grössere Softwareprojekte aus anderen Anwendungsgebieten repräsentativ zu sein.

Ähnliche Ziele verfolgen zwei Gruppierungen innerhalb der GI: der Arbeitskreis "Management von Softwareprojekten" innerhalb der Fachgruppe 2.1.1 "Software Engineering", der seit 1993 tätig ist [Elz92] und die Fachgruppe 5.1.2 "Projektmanagement" im Fachausschuß 5.1. Beide Gruppen kooperieren eng, möchten aber durch die Beibehaltung ihrer ursprünglichen Anbindung den notwendigen interdisziplinären Charakter der Arbeit hervorheben. Eine ausführliche Darstellung der Ziele und der Arbeit des erstgenannten AK findet sich in [Elz94]. Aus den Vorträgen und Diskussionen in den Workshops und Arbeitskreisen ergibt sich etwa folgendes Zustandsbild, das sich in etwa auch mit der "Literaturlage" und dem größten Teil der beruflichen Erfahrungen des Verfassers deckt.

3 Derzeitiger Zustand

Daß zwischen Projektgröße, Organisationsformen und eingesetzter Technik enge Zusammenhänge bestehen, und deshalb "Patentlösungen" mit Hilfe bestimmter einzelner Verfahren nicht möglich sind, scheint heute weitgehend verstanden zu sein. Diese Einsicht wurde von Pionieren zwar schon früh vertreten [Bro75, Bro87], hat sich aber nur langsam durchgesetzt. Sie wurde vermutlich von den meisten betroffenen Praktikern deswegen nicht akzeptiert, weil diese je nach ihrer beruflichen Vergangenheit entweder nur die technischen Aspekte sahen oder aber die organisatorischen Probleme mit Methoden des traditionellen Industriemanagement lösen zu können glaubten.

Daß professionelles Projektmanagement für die erfolgreiche Entwicklung von Software nichttrivialen Umfangs unumgänglich ist, ist inzwischen auch allgemein anerkannt. Es scheint aber, daß das notwendige "handwerkliche Wissen" und die praktische Übung in einfachsten Managementtechniken noch nicht im nötigen Maß Allgemeingut sind. Dies geht beispielsweise aus einer interessanten Untersuchung hervor, im Rahmen derer festgestellt wurde, daß Firmen, die den Übergang von Stufe 1 zu Stufe 2 des CMM [Hum89] mehrfach nicht schafften, speziell Defizite bei den einfachsten Projektleitungstechniken hatten: Projektplanung und Projektverfolgung und -überwachung [HZG97]. Der Verfasser kann diesen Eindruck auf Grund eigener Beobachtungen bestätigen.

Besonders erfolgreich scheint die breite Durchsetzung der Einsicht zu sein, daß Softwareentwicklung als Prozeß verstanden werden muß, der nur mit Hilfe einer ganzen Palette von Maßnahmen verbessert werden kann. Außerdem muß an dieser Verbesserung laufend gearbeitet werden. Dieser vor allem durch das CMM ausgelöste Ansatz des "Software Process Improvement" (SPI) wird zwar von manchen Institutionen auch wieder im Sinne einer "silbernen Kugel" [Bro87] vermarktet, trägt aber doch wesentlich weiter als die früheren eindimensionalen Methoden.

Ganz unumstritten scheint er aber nicht zu sein. Der Verfasser kennt aus eigener Erfahrung die Bedenken selbst eingeführter Firmen aus dem Verteidigungsbereich in Bezug auf zu viel Bürokratismus, der durch das Modell erzwungen würde. Weitere Hinweise finden sich auch in der Literatur (z.B. in [Bac94]). Selbst in einem insgesamt sehr positiven Erfahrungsbericht, wie ihn beispielsweise [HZG97] darstellt, werden gewisse Schwierigkeiten, die kleine Firmen im zivilen Bereich mit dem CMM haben könnten, nicht ausgeschlossen.

Daß andererseits selbst durch außerordentlich gut durchentwickelte formale Qualitätssicherungsverfahren, wie sie beispielsweise in Raumfahrtagenturen notwendig sind, schwere Fehlleistungen nicht verhindert werden können, zeigt der Verlust des ersten Exemplars der ARIANE 5 Rakete. Die in einem offiziellen Bericht [LIO96] dargestellte Fehlerursache scheint dem Verfasser aber eher auf einem leider weit verbreiteten "Schnellschußverfahren" bei Managemententscheidungen zu beruhen als auf einem softwaretechnischen Problem. Der dargestellte Versuch, durch die Wiederverwendung von Software den ganzen Testaufbau für ein Subsystem einzusparen, erinnert ihn stark an ähnliche Erlebnisse an früheren Arbeitsplätzen. Andererseits ist er ein starker Hinweis auf die Gefahren, die in Zukunft auf die Projektverantwortlichen durch den Einsatz ungeprüfter oder nicht prüfbarer Standardkomponenten zukommen können.

Besonders vielversprechend erscheint dem Verfasser deshalb eine - richtig verstandene - Anwendung der in "ISO 900x" enthaltenen Gedanken (beispielsweise in [DIN91]). Dieses Normenwerk wurde zwar zunächst nicht für Software, sondern ganz allgemein für die Qualitätssicherung von Produkten und Dienstleistungen aller Art entwickelt, regt aber zu einer Denkweise in vermaschten Wirkungskreisen - in Netzwerken also - an, die für die Softwareentwicklung der Zukunft tragfähiger erscheint als manche speziell für Software gedachte Vorgehensweise. Bild 1 soll dies illustrieren.

Die Forderungen in ISO 900x erzwingen es außerdem, alle einen Entwicklungsprozeß betreffenden Entscheidungen bewußt zu treffen und zu dokumentieren. Dabei haben aber andererseits die Verantwortlichen die volle Freiheit, welche Verfahren oder Organisationsformen sie in Bezug auf den einzelnen Geschäftsfall anwenden wollen. Damit ist ISO 900x insbesondere unabhängig von einem konkreten Phasenmodell. Als positiv ist weiterhin zu werten, daß die Verantwortung der obersten Leitung für die Qualität festgelegt wird und daß die Wichtigkeit organisatorischer Maßnahmen, wie z.B. Reviews, herausgearbeitet wird. Weiterhin ist wichtig, daß Konfigurationsmanagement wieder in das Bewußtsein der Verantwortlichen gerückt und explizit auf Testverfahren hingewiesen wird. Schließlich wird ausdrücklich gefordert, daß frühere Designerfahrungen verwendet werden. Dies ist eine starke Motivation zur Einführung planmäßiger Wiederverwendung von Software.

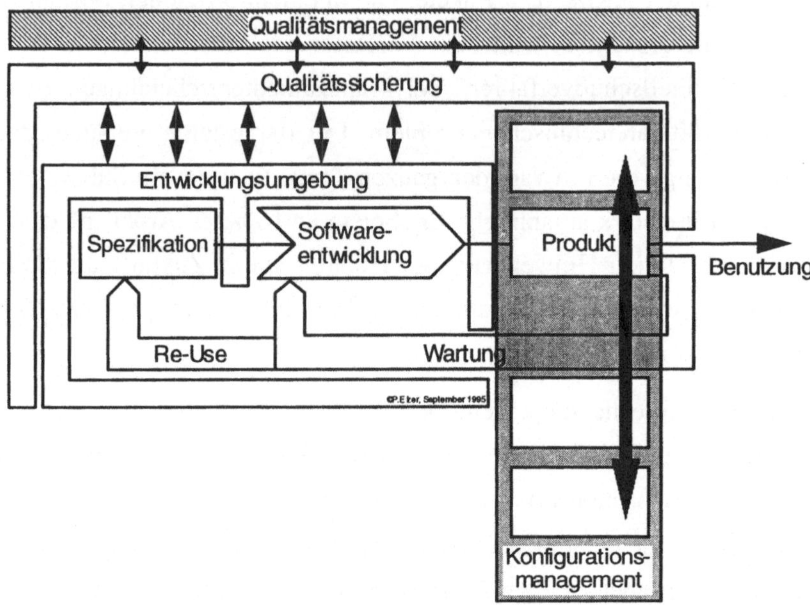

Bild 1: Das Netzwerk des Entwicklungsprozesses gemäß ISO 900x

Trotz des gezeichneten eher positiven Bildes des gegenwärtigen Zustandes der Softwareentwicklung sollte aber doch erwähnt werden, daß immer noch zu wenig echtes empirisches Material in Bezug auf die Wirksamkeit einzelner Entwicklungshilfsmittel oder Managementmethoden vorliegt. Die meisten empirischen Untersuchungen beschäftigen sich nur mit der nachträglichen Überprüfung des Einsatznutzens vorher definierter normativer Modelle. Ein Hinweis darauf, daß wirklich breit angelegte und unvoreingenommene Untersuchungen [BrF94] zu unerwarteten Ergebnissen kommen können, findet sich in [HSF96]: in dieser sehr detaillierten statistischen Auswertung des in einer großen Zahl von Entwicklungsprojekten gewonnenen umfangreichen Materials finden sich Hinweise darauf, daß Benutzerbeteiligung bei der Softwareentwicklung eher kontraproduktiv sein könnte. Dies stünde im Widerspruch zu einer über Jahre hinweg allgemein für wahr gehaltenen Annahme. Nach den "Gesetzen der allgemeinen Lebenserfahrung" ist eigentlich zu erwarten, daß weitere empirische Untersuchungen noch mehr Widersprüche zwischen "common folklore" und tatsächlichen Gegebenheiten aufdecken würden.

Zusammenfassend kann man sagen, daß die Diskussion über die beherrschenden Themen der vergangenen zwei Jahrzehnte, wie etwa

-Entwicklungswerkzeuge

-projektorientierte Lebensdauerzyklusmodelle

-Kostenschätzung und -kontrolle

-Qualitätsicherung

-Kreativität und Motivation von Mitarbeitern

durchaus Früchte getragen hat. Den Entwicklungsverantwortlichen steht heute ein ausreichend großes Repertoire an Lösungsansätzen zur Verfügung - wenn sie diese zur Kenntnis nehmen und zu nutzen verstehen.

Es scheinen sich aber neue Veränderungen anzubahnen, die nach Ansicht des Verfassers in der "offiziellen" Diskussion noch nicht ausreichend berücksichtigt werden. Dieser Eindruck wurde hauptsächlich durch zwei Beobachtungen erweckt.

4 Beobachtete Tendenzen

Erstens: Auswertungen von bei den genannten IFAC-Workshops gemachten (anonymen) Umfragen ergaben zunächst, daß die Produktivität von Entwicklern (zumindest bei Echtzeitanwendungen) von Beginn der 70-er bis Ende der 80-er Jahre trotz aller technischen Weiterentwicklungen praktisch unverändert geblieben zu sein schien. Anfang der 90-er Jahre wurden die Werte plötzlich nicht mehr vernünftig interpretierbar. Der reine Mittelwert der Produktivität stieg sprunghaft an, gleichzeitig wuchs aber trotz etwa gleichbleibender Stichprobengröße die Streuung so stark, daß keine einheitliche Deutung mehr möglich war. Dieser Eindruck wurde durch Beobachtungen in einer großen Firma bestätigt, die nach der - etwa im gleichen Zeitraum durchgeführten - Einführung eines anerkannten Schätzverfahrens schlechtere Planungsdaten erhielt als vorher. An der Professionalität der Einführung und Verwendung des Schätzverfahrens konnte es nicht gelegen haben.

Zweitens: Schon während seiner Tätigkeit in der Industrie (Bau großer Leitsysteme) mußte der Verfasser mehrfach miterleben, wie durch plötzliche Änderungen der Marktpolitik von Zulieferern Projekte - oft kurz vor ihrem eigentlich erfolgreichen Abschluß - eingestellt werden mußten oder zumindest finanzielle Fehlschläge wurden. Zunächst betraf das nur Hardware, Betriebssysteme, Compiler, etc. und konnte durch den Übergang zu "UNIX" und gewisse "strategische Partnerschaften" für einige Jahre aufgefangen werden. Dies gelang um so eher, als die eigene Firma durch ihre schiere Größe als Vertragspartner fast immer in einer guten Position war. Gegen Ende der 80-er Jahre trat das Problem aber dann auch auf Gebieten auf, von denen man vorher angenommen hatte, sie völlig zu beherrschen, wie etwa spezielle Grafikpakete für automatisierungstechnische Anwendungen. Es wurde verschärft dadurch, daß es hier auch - wegen der in den einzelnen Geschäftsfeldern unterschiedlichen Anforderungen und Zeitpläne - nicht mehr ohne weiteres möglich war, die Größe der Gesamtfirma ins Spiel zu bringen.

Als Leiter eines kleinen Instituts einer kleinen Universität muß der Verfasser nun seit 1990 in aller Schärfe miterleben, wie die eigentliche "Produktivität", d.h. die Entwicklung forschungsrelevanter Software (auf den Gebieten "Mensch-Maschine-Schnittstelle", "Virtual Reality" und Simulation) oder der Betrieb umfangreicher Praktika (Softwareentwicklung in der Automatisierungstechnik) dadurch nahezu "auf Null" reduziert wird, daß man gezwungen ist, jeweils neue Versionen von "Standardsoftwarepaketen" wieder zusammen funktionsfähig zu machen. Selbst das in einer industriellen Umgebung bewährte "Einfrieren" von Versionen für die Lebensdauer einer Produktlinie ist kaum anwendbar. Spätestens die Ersatzbeschaffung eines nicht mehr reparablen Rechners zieht den Neuaufbau der gesamten Softwareumgebung nach sich, was im schlimmsten Fall dazu führen kann, daß teure Anwendungspakete nachlizenziert werden müssen. Umgekehrt zwingt der rasant steigende Bedarf an Rechnerleistung, den "neue" Software - bei eigentlich gleicher Funktionalität - jeweils aufweist, zur Anschaffung immer leistungsfähigerer Rechner, was wiederum den erhofften "Preisverfall" meist mehr als kompensiert.

Nun haben Universitätsinstitute den "Wettbewerbsvorteil", daß es immer wieder gelingt, Sonderpreise, Campuslizenzen oder Spenden auszuhandeln. Eine kleine bis mittelständische Firma hat diese Möglichkeiten aber nicht! Trotzdem bleibt auch Instituten das Problem der direkten und indirekten Umstellungskosten einschließlich der Umschulung von Mitarbeitern. Nicht zu unterschätzen sind auch die Kosten der Umstellung von Datenbeständen bei Änderungen von Text- und Grafiksystemen sowie von Verwaltungssoftware oder die ihrer Rekonstruktion bei Verlust. Viel Arbeitsaufwand verursachen schließlich Inkompatibilitäten bei Schnittstellen zwischen verschiedenen Standardpaketen, ja selbst zwischen verschiedenen Versionen des gleichen Standardpakets auf verschiedenen Rechnern.

Unabhängig von diesen Schwierigkeiten ist ein neues Problem für Projektleiter in der Praxis durch eine an und für sich begrüßenswerte Entwicklung entstanden: die Verbreitung von "Managementverfahren" und "-werkzeugen". Unabhängig von einer Bewertung ihres prinzipiellen Einsatznutzens verursachen sie die gleichen Probleme der Schulungs- und Umstellungskosten wie andere Standardpakete auch.

5 Mögliche Folgerungen

Die genannten Beobachtungen wurden dem Verfasser von verschiedensten Seiten bestätigt. In weiten Bereichen der Anwendungsentwicklung ist also ein neues Verständnis des Entwicklungsvorganges nötig: Man entwickelt Systeme nicht mehr durch Programmieren, sondern durch Zusammensetzen von Standardpaketen mit eventuellen Ergänzungen. Der "klassische" Entwicklungsprozeß wird nicht aussterben, aber wohl nur noch in Firmen stattfinden, die diese Pakete entwickeln.

Dies wird vor allem Konsequenzen für die notwendige Personalqualifikation haben. Dabei sollen zunächst die Entwickler betrachtet werden:

Die genannten - meist klein erscheinenden - Ergänzungen erfordern häufig noch mehr technisches Verständnis und ein umfangreicheres Fachwissen als es bei der Programmierung notwendig war. Vor allem ist es nötig, sich in alle verwendeten Komponenten einarbeiten zu können. Diese Situation ist vergleichbar mit der bei der "klassischen" Wartung von Programmsystemen, nur daß es sich grundsätzlich um mehrere Systeme handelt, die sich außerdem von Release zu Release ändern können, ohne daß man selbst Einfluß darauf hat. In Anbetracht der alten Diskussion um die nötige Qualifikation von Wartungsteams sei hier auf die dabei genannten psychologischen Aspekte hingewiesen. Für Anwendungsentwickler werden vermutlich in Zukunft weniger Methodenkompetenz, mathematisch orientierte Problemlösungskompetenz oder Kreativität im Detail gefragt sein als tiefes technisches Verständnis, gekoppelt mit rascher Auffassungsgabe, "analytischem Spürsinn" und der Fähigkeit, sich schnell in komplexe neue Systeme einzuarbeiten und - vor allem - Trends richtig zu erkennen und einzuschätzen.

Diese letztgenannte Fähigkeit wird auch für Projektverantwortliche noch wichtiger werden [Hoc91]. Außerdem werden sie sich in Zukunft noch mehr nach außen als nach innen orientieren müssen. Gesichtspunkte der Vertragsgestaltung mit Lieferanten werden noch wichtiger werden. Außerdem müssen schon bei der Planung von Projekten Alternativen für den Fall von Änderungen der Zuliefersituation vorgesehen werden. Insgesamt gesehen hat man den Entwicklungsprozeß weniger unter Kontrolle als bisher. Also werden Risikoabschätzungen noch wichtiger werden.

Bisher wurde SW-Management vor allem unter dem Gesichtspunkt der Neuentwicklung gesehen. Hierfür ist, wie eingangs erwähnt, die Strukturierung in einem "Projekt" der geeignetste Weg. Diese Organisationsform war aber in der traditionellen Industrie wenig verbreitet. Der Verfasser hat selbst miterlebt, welche Schwierigkeiten die geistige und organisatorische Umstellung auf die mit dem "Projektmanagement" verbundenen Strukturen und Vorgehensweisen größeren Firmen intern bereitete. Es erscheint deshalb selbstverständlich, daß die jetzt anstehende Umorientierung erneut Schwierigkeiten mit sich bringen und deshalb vielleicht wieder als "Krise" empfunden werden wird. Dabei kann es sehr hilfreich sein, altbekannte Vorgehensweisen, wie das "Konfigurationsmanagement" [Pen95] auf geeignete Weise mit dem "Projektmanagement" zu verbinden.

Die Kostenschätzung wird nicht einfacher werden. Zunächst erschien es so, daß durch die - eigentlich - feststehenden Preise die immer erhebliche Schätzunsicherheit beim klassischen Entwicklungsprozeß verringert würde. Dafür entstehen neue Unsicherheiten durch die Preispolitik der Zulieferer, die diese ja oft nach Kriterien völlig fremder Märkte gestalten müssen. Außerdem verlieren aus obengenannten Gründen die bisher gebräuchlichsten Produktivitätsmaße für die Arbeit von Entwicklern ihre Gültigkeit. Es wird also nötig sein, neue Kostenschätzungsverfahren zu erarbeiten.

Es entstehen neue Probleme der Qualität: man hat die Entwicklung nur noch zu einem kleinen Teil im Griff und muß lernen, die Qualität von - teilweise extrem komplexen - Zulieferteilen einzuschätzen. Es erscheint nur als ein erster - wenn auch notwendiger - Schritt, genau auf die Zertifizierung von Zulieferern zu achten. Kaum durchsetzbar scheint es zu sein, über eine geeignete Vertragsgestaltung bei Zulieferern die eigenen QS-Prozeduren durchzusetzen. Diese sind ja, wie eingangs erwähnt, meist größer und marktmächtiger als die eigene Firma oder Organisationseinheit.

Es erscheint sinnvoll, dieses Problem als einen Spezialfall der Wiederverwendung von Software zu betrachten. Dafür wird es aber nötig sein, Ansätze zu entwickeln, wie die Dokumentation von Software besser, problemorientierter, lesbarer gestaltet werden kann [Elz91b]. Vor allem wird es notwendig sein, umfangreiche Dokumentationen so aufzubereiten daß die Leser in ihnen leicht und treffsicher navigieren können [ElK97]. Andererseits muß weiter daran gearbeitet werden, die Vorgehensweise bei der Spezifikation von Software zu verbessern. Dadurch werden zwar katastrophale Ereignisse, wie der schon genannte Verlust der ARIANE 5 nie völlig ausgeschlossen werden können, aber durch die bessere Offenlegung von Entwurfsentscheidungen wird es auch leichter sein, kritische Parameter in zu verwendenden Standardkomponenten zu identifizieren und ihren Einfluß zu beurteilen.

Die Verfahren zur Verbesserung des Softwareentwicklungsprozesses müssen weiterentwickelt werden. Nach Ansicht des Verfassers weist selbst CMM noch zu viele Züge der normativen Modelle alter Art auf. Eine neue Art der Projektabwicklung ist im Enstehen, also sollte von vornherein darauf geachtet werden, rechtzeitig und unvoreingenommen empirische Daten zu erheben.

Außerdem wird über Prozeßmodelle ganz allgemein - immer noch - zu stark technologiezentriert und projektorientiert argumentiert. Selbst ISO 900x, das der Verfasser wegen seiner Flexibilität und Methodenfreiheit als einen "Schritt in die richtige Richtung" betrachtet, behandelt die Entwicklung hauptsächlich unter dem Projektgedanken. Die oben erwähnte Kombination von Projekt- und Konfigurationsmanagement sollte schon im Prozeßmodell angelegt sein.

Nach all diesen Gedanken in Bezug auf mögliche Anpassungen des Denkens über Projektmanagement an eine neue Situation ist es aber auch sicher angebracht zu fragen: was bleibt von den bisherigen Einsichten über gutes Projektmanagement?

Nach Ansicht des Verfassers bleibt sehr viel. Zunächst ist festzustellen, daß es die Softwareentwicklungsprojekte der klassischen Art auch weiterhin geben wird. Es wird immer wieder neue Anwendungen geben, und die Komponentenhersteller sind sicher gut beraten, ihre Arbeit durch die Anwendung aller bisher bekannten Verfahren optimal zu gestalten. Aber auch für die dargestellten "Projekte neuer Art" werden Aspekte wie Multidimensionalität der Vorgehensweise, gute Menschenführung, ausgewogene Einschätzung der Leistungsfähigkeit von Softwarewerkzeugen, dem Projekt angemessene Abwicklungsmodelle, etc. weiterhin von großer Bedeutung sein.

Literatur

[Bac94] J. Bach: Enough about process: what we need are heroes, IEEE Software, Vol. 12, No.2, 1994, S. 96-98

[BrF94] F. C. Brodbeck, M. Frese (Hrsg.): Produktivität und Qualität in Softwareprojekten, R. Oldenbourgh Verlag, München, Wien, 1994

[Bro75] F. P. Brooks, Jr.: The Mythical Man Month, Addison Wesley, 1975

[Bro87] F. P. Brooks: No silver bullet: Essence and accidents of software engineering, IEEE Computers, Vol. 20, No. 4, 1987, S. 10-19

[BuR70] J. N. Buxton, B. Randell: Software engineering Techniques, Conference Report, NATO Science Committee, Brüssel, 1970

[DIN91] DIN ISO 9000-3: Qualitätsmanagement- und Qualitätssicherungsnormen, Leitfaden für die Anwendung von ISO 9001 auf die Entwicklung, Lieferung und Wartung von Software, Beuth Verlag, Berlin, Köln, 1991

[ElH92] P. Elzer, V. Haase (Hrsg.): Proceedings of the 4th IFAC Workshop on Experience with the Management of Software Projects, Seggau, Österreich, Mai 1992, Pergamon Press, Oxford, 1992

[ElK97] P. Elzer, U. Krohn: Visualisierung zur Suche in komplexen Datenbeständen, angenommener Vortrag bei der HIM´97, Dortmund, 1997

[ElR96] P. Elzer, R. Richter (Hrsg.): Proceedings of the 5th IFAC Workshop on Experience with the Management of Software Projects, Karlsruhe, September 1995, Pergamon Press, Oxford, 1996

[Elz87] P. Elzer (Hrsg.): Experience with the Management of Software Projects, Proceedings of the IFAC/IFIP Workshop, Heidelberg 1986, Pergamon Press, Oxford, 1987

[Elz90] P.Elzer: Aspekte der Menschenführung bei der Abwicklung von Software-Projekten, GI-Workshop "Arbeitsverfahren in der Softwareentwicklung", Königswinter, Mai 1990

[Elz91a] P. Elzer (Hrsg.): Multidimensionales Software - Projektmanagement, ein Handbuch für Projektleiter zur Entwicklung komplexer Informationssysteme, AIT Verlags GmbH, Hallbergmoos, 1991

[Elz91b] P. Elzer: Reuse, a Problem of "Understanding" Designs; in: R. Prieto-Diaz, W. Schäfer, J. Cramer, S. Wolf (Hrsg.): Proceedings of the First International Workshop on Software Reusabilty, FB Informatik, Universität Dortmund, Juni 1991

[Elz92] P. Elzer: Aufruf zur Mitarbeit, Softwaretechnik-Trends, Band 12, Heft 3, 1992

[Elz94] P. Elzer: Bericht über die Aktivitäten des Arbeitskreises "Management von Softwareprojekten", in: Tagungsband des Workshop der GI Fachgruppe 5.1.2 "DV-Projektmanagement, Erfahrungsberichte und neue Ansätze", München, Juni 1994

[Gla97] R. L. Glass: Revisiting the Industry/Academe Communication Chasm, CACM, Vol. 40, No.6, 1997, S. 11-13

[Hoc91] D. J. Hoch: Management von Technologie-Diskontinuitäten in Informatik-Projekten, in: [Elz91a], S. 153-179

[HSF96] T. Heinbokel, S. Sonnentag, M. Frese, W. Stolte, F. C. Brodbeck: Don´t underestimate the problems of user centredness in software development projects - there are many! BEHAVIOUR AND INFORMATION TECHNOLOGY, Vol. 15, No. 4, 1996, S. 226-236

[Hum89] W. S. Humphrey: Managing the Software Process, Addison-Wesley, 1989

[HZG97] J. Herbsleb, D. Zubrow, D. Goldenson, W. Hayes, M. Paulk: Software quality and the capability maturity model, CACM, Vol. 40, No.6, 1997, S. 30-40

[Küc91] A. Küchle: Wirtschaftliche Effizienz von Softwarewerkzeugen, in: [Elz91a], S. 113-151

[LIO96] J. L. Lions (Hrsg.): ARIANE 5 - Flight 501 Failure, Report by the Inquiry Board, ESA, Paris, 1996

[MiE88] R. Milovanovic, P. Elzer, (Hrsg.): Proceedings of the 2nd IFAC Workshop on Experience with the Management of Software Projects, Sarajevo, September 1988, Pergamon Press, Oxford, 1988

[MoE89] F. Mowle, P. Elzer, (Hrsg.): Proceedings of the 3rd IFAC Workshop on Experience with the Management of Software Projects, Lafayette, USA, October 1989, Pergamon Press, Oxford, 1989

[Pen95] G. Penzenauer: Change and Configuration Management (CCM) in heterogeneous Environment, in: [ElR96], S. 101-105

[Rem87] U. Rembold (Hrsg.): Einführung in die Informatik für Naturwissenschaftler und Ingenieure, Hanser Verlag, München, Wien, 1987

Systematische Aufbereitung und Nutzung von Erfahrungen aus Softwareprojekten bei Daimler-Benz

Dieter Landes, Kurt Schneider

Abstract

Die Kompetenz, effizient qualitativ hochwertige Software selbst entwickeln zu können oder von anderen entwickeln lassen zu können, beruht wesentlich auf der Fähigkeit, systematisch aus Erfahrungen zu lernen, die in diesem Umfeld bereits gemacht wurden. In diesem Beitrag wird ein Projekt vorgestellt, in dem bei der Daimler-Benz AG eine sog. Experience Factory etabliert wird, die eine systematische Aufzeichnung und Nutzung von Erfahrungen ermöglicht. Das in der Literatur beschriebene Konzept der Experience Factory muß in diesem Projekt zunächst auf die konkreten Gegebenheiten angepaßt und erweitert werden. Erstes Anwendungsfeld ist die Aufbereitung und Nutzung konkreter Erfahrungen zum Qualitätsmanagement bei der Entwicklung administrativer Software.

1 Einführung

In den letzten Jahren ist die Bedeutung der Software für ein Unternehmen stetig gewachsen: so werden beispielsweise die Eigenschaften hergestellter Produkte in zunehmenden Maß durch Software bestimmt oder Geschäftsprozesse mehr und mehr durch Softwarewerkzeuge unterstützt. Dies hat zur Folge, daß eine Steigerung der Produktivität und Qualität im Software Engineering mehr und mehr zum kritischen Erfolgsfaktor für Unternehmen wird, was sich auch an der zunehmenden Verbreitung von Ansätzen zur Bewertung und Verbesserung von Softwareprozessen wie z.B. ISO9000 ff. [KeJ96], CMM [PCC93, PWC95], Bootstrap [KuB94], SPICE [Rou95] oder QIP/GQM [Bas85, Bas93, BCR94a] ablesen läßt.

Einen wesentlichen Beitrag zur Produktivitätssteigerung leistet die Fähigkeit, Erfahrungen, die innerhalb eines Softwareprojekts gemacht werden, in ähnlich gelagerten Projekten systematisch wiederverwenden zu können. Ein Konzept, das ein systematisches Lernen aus Erfahrungen zum Ziel hat, ist die sogenannte Experience Factory [Bas93, BCR94b], mit der vor allem das Software Engineering Laboratory der NASA seit Jahren erfolgreich arbeitet [BCM92].

Die Daimler-Benz AG (DBAG) hat im Jahr 1996 drei Projekte gestartet, in denen jeweils eine Experience Factory in einem Bereich des Unternehmens etabliert wird. Eines dieser Projekte (*VP-Admin*) soll in diesem Beitrag vorgestellt werden. Ziel von *VP-Admin* ist neben dem Aufbau einer konkreten Experience Factory, die zunächst Erfahrungen zum Themenbereich Qualitätsmanagement für administrative Software umfassen wird, das Sammeln von Erfahrungen, wie bei der Etablierung einer Experience Factory vorzugehen ist und welche Anpassungen des in der Literatur beschriebenen Konzepts an das gegebene Umfeld erforderlich sind[1].

Im folgenden Abschnitt wird zunächst der theoretische Hintergrund des Experience-Factory-Konzepts erläutert, bevor in Abschnitt 3 die Erweiterungen und Anpassungen diskutiert werden, die in *VP-Admin* erforderlich sind. In Abschnitt 4 wird auf die Etablierung einer konkreten Experience Factory im Rahmen von *VP-Admin* und deren inhaltliche Ausrichtung eingegangen. Abschnitt 5 gibt eine kurze Zusammenfassung.

2 Die Experience Factory nach Basili et al.

Das Konzept der Experience Factory ist eng verknüpft mit dem von Basili et al. entwickelten *Quality Improvement Paradigm* (QIP) [Bas85, Bas93], das einen

---

[1] *VP-Admin* beinhaltet darüber hinaus ein weiteres Teilprojekt, das sich mit dem Themenfeld Produktivitätssteigerung durch Wiederverwendung objekt-orientierter Komponenten und Fragen der Software-Architektur beschäftigt. Auf dieses Teilprojekt soll im vorliegenden Beitrag jedoch nicht weiter eingegangen werden.

Rahmen für die systematische Verbesserung von Softwareprozessen und der Qualität der dabei entwickelten Produkte darstellt. QIP setzt sich aus sechs Einzelschritten zusammen, die zyklisch durchlaufen werden sollten:

1. Charakterisierung des Softwareprojekts und seines Umfelds
2. Festlegung quantifizierbarer Verbesserungsziele
3. Auswahl der zu betrachtenden Teilprozesse und Festlegung geeigneter Maßzahlen in Abhängigkeit von den zugrundegelegten Verbesserungszielen
4. Durchführung der Prozesse und Erhebung der festgelegten Maßzahlen
5. Interpretation der Maßzahlen und Ableitung von Verbesserungsmaßnahmen
6. Aufbereitung und Bereitstellung gewonnener Erfahrungen

Die Aufgaben der Experience Factory leiten sich aus dem letzten Punkt ab: sie ist dafür zuständig, Erfahrungen aus einem Verbesserungsprogramm zu erfassen und aufzubereiten und Rückkopplungen an die Entwicklungsprojekte zu geben. Die Experience Factory selbst ist also eine organisatorische Einheit, die Daten und Informationen, die in Entwicklungsprojekten gewonnen werden, analysiert und interpretiert und sie ggf. so aufbereitet, daß sie für ähnlich gelagerte Projekte wiederverwendbar sind. Die gewonnenen Erfahrungen werden in einem „Erfahrungsspeicher", der sogenannten Experience Base, abgelegt. Ein entscheidender Aspekt ist jedoch, daß die Experience Factory mehr ist als ein mehr oder minder intelligenter Erfahrungsspeicher: sie ist auch dafür verantwortlich, Entwicklungsprojekte aktiv zu unterstützen, d.h. für diese Projekte relevante Erfahrungen zu identifizieren und bereitzustellen und die Umsetzung dieser Erfahrungen beratend zu begleiten. Der Zusammenhang zwischen Experience Factory und Projektorganisation ist in Abb. 1 nochmals wiedergegeben.

Durch eine Experience Factory wird also erreicht, daß Erfahrungen explizit gemacht und personenunabhängig abgelegt werden, d.h. der Erfolg von Entwicklungsprojekten weniger stark vom individuellen Know-How einzelner Projektbeteiligter bestimmt wird. Die organisatorische Trennung von Projektorganisation und Experience Factory sorgt dafür, daß die Projektorganisation nicht damit belastet wird, unter dem üblicherweise herrschenden Projektdruck auch noch Aufwände für die Aufzeichnung und Aufbereitung von Erfahrungen erbringen zu müssen, die in erster Linie anderen Projekten zugute kommen.

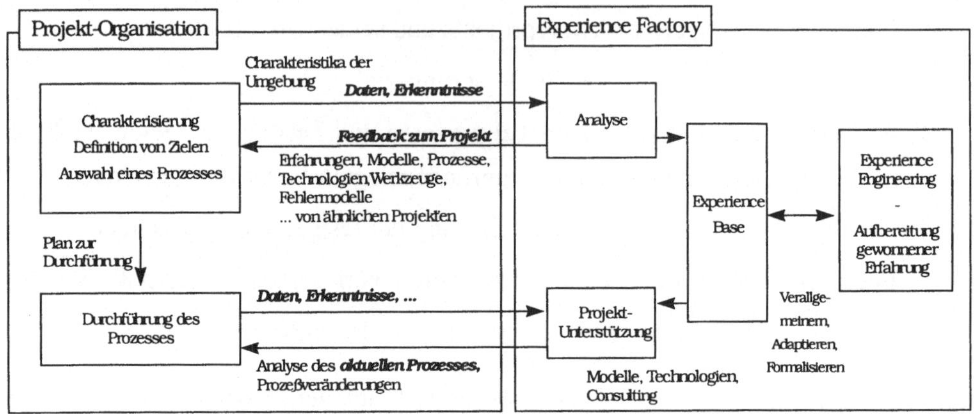

Abbildung 1: Interaktion zwischen Projektorganisation und Experience Factory

3 Anpassungen und Erweiterungen

Das Konzept der Experience Factory, wie sie im Software Engineering Laboratory der NASA praktiziert wird [BCM92], kann wegen des unterschiedlichen Umfelds nicht ohne weiteres übernommen werden. Im folgenden werden einige Aspekte aufgeführt, die im Rahmen von *VP-Admin* näher zu untersuchen sind und bezüglich derer das Konzept der Experience Factory zu verfeinern ist.

3.1 Strukturierung von Informationen

Basili et al. gehen in ihren Arbeiten nur sehr allgemein auf die Inhalte der Experience Base ein. So wird z.B. in [BCR94b] von Modellen unterschiedlicher Ausprägungen, Meßwerten, Produkten und anderen Formen von Wissen gesprochen; offen bleibt jedoch, wie diese Informationen zu strukturieren und zueinander in Bezug zu setzen sind, damit sie im Bedarfsfall effizient lokalisiert und wiederverwendet werden kann.

Zu dieser Frage wurden in einem Projekt unter Beteiligung der DBAG bereits Vorarbeiten geleistet: als Beschreibungsmittel für Erfahrungspakete, also Beobachtungen, die über den konkreten Einzelfall hinaus gültig sind, wurden in Anlehnung an die Design Patterns im objekt-orientierten Umfeld [GHJ94] sog. Quality Patterns vorgeschlagen und erprobt [HoK97]. Quality Patterns bauen auf dem von Minto vorgeschlagenen Pyramidenprinzip [Min87] zur Strukturierung von Information auf und setzen sich aus drei Grundbestandteilen zusammen:

- einem Klassifikationsteil, in dem ein Erfahrungspaket anhand bestimmter Attribute kategorisiert wird,
- einem Teil, der eine Kurzfassung der gemachten Erfahrung, eine Beschreibung des zugrundeliegenden Problems und seiner Lösung sowie des Kontexts, in dem die Erfahrung gewonnen wurde, enthält sowie
- Zusatz- und Verwaltungsinformation wie etwa Beispiele, weitere Erläuterungen, Querverweise zu anderen Erfahrungspakten u.ä.

Nähere Erläuterungen der einzelnen Komponenten und einige Beispiele von Quality Patterns finden sich in [HoK97].

Erfahrungspakete werden in *VP-Admin* mit Hilfe von Quality Patterns beschrieben, um zu klären, ob bzw. wie gut sie sich als Strukturierungsprimitv auch in diesem Umfeld bewähren.

Allerdings sind Quality Patterns für sich allein genommen sicher noch nicht ausreichend, da die Informationen und Eindrücke, die in Projekten gesammelt werden, erst analysisert, verdichtet und abstrahiert werden müssen, bevor sie als Erfahrungspaket mit Hilfe eines Quality Patterns beschrieben werden können. Die Experience Base enthält folglich neben den eigentlichen Erfahrungspaketen eine Vielzahl von Informationsbruchstücken wie etwa Interviewmitschriften, Tätigkeitsberichten, Hintergrundberichten, Analyseergebnissen u.ä., die sich aufeinander beziehen, voneinander abgeleitet, einander widersprechen oder in weiteren Beziehungen zueinander stehen können. Diese Querbezüge müssen geeignet beschrieben werden. Als Ausgangspunkt für die Definition eines angemessenen Vokabulariums von Beziehungstypen zur Beschreibung dieser Querbezüge erscheinen Modelle geeignet, wie sie zur Erfassung von Design Rationales entwickelt wurden

[MoC96]. Welches der vorgeschlagenen Modelle im Umfeld der Experience Factory das geeignetste und wie es ggf. zu adaptieren ist, ist im Rahmen von *VP-Admin* noch zu klären.

3.2 Einbindung der Mitarbeiter der Projektorganisation

Neben der Frage, wie die Informationen in der Experience Base zu strukturieren sind, ist eine weitere Fragestellung die, wie die Mitarbeiter aus der Projektorganisation bei der Formulierung von Erfahrungspaketen einzubinden sind.

Einerseits läßt sich die Experience Factory als Instrument innerhalb einer Lernenden Organisation nach Senge [Sen90] ansehen, der als Merkmale einer solchen Organisation u.a. Kontinuität des Lernens nennt. Das hat zur Folge, daß potentielle Erfahrungspakete idealerweise sofort festgehalten werden sollten, wenn die entsprechenden Erkenntnisse im Entwicklungsprojekt gewonnen werden. Es muß also die Möglichkeit bestehen, daß die Mitarbeiter der Projektorganisation selbst Information in die Experience Base eingeben können.

Andererseits stecken viele relevante Informationen implizit in den Köpfen der Mitarbeiter eines Entwicklungsprojekts, ohne daß sie in der Lage wären, diese explizit zu formulieren. Daher muß innerhalb der Experience Factory aber auch eine Rolle analog zu der des Knowledge Engineers bei der Wissensakquisition für wissensbasierte Systeme ausgefüllt werden, um an diese implizite Information heranzukommen und die erforderliche Abstraktion der Information und ihre Einordnung in den Gesamtzusammenhang innerhalb der Experience Base vorzunehmen.

In *VP-Admin* wird an Lösungsansätzen gearbeitet, wie diese scheinbar gegensätzlichen Forderungen miteinander in Einklang gebracht werden können.

3.3 Suche nach relevanten Erfahrungen

Aus Sicht von Basili et al. erfolgt die Identifikation relevanter Erfahrungen zur erneuten Verwendung in einem neuen Entwicklungsprojekt ausschließlich durch die

Projektunterstützung innerhalb der Experience Factory (vgl. Abbildung 1). Damit ist aber wieder eine starke Abhängigkeit von Einzelpersonen gegeben, wie sie etwa auch die Ebene 1 des CMM-Reifegradmodells [PCC93, PWC95] charakterisiert: um eine Situation zu erkennen, in der für ein laufendes Projekt die Nutzung ganz bestimmter Erfahrungspakete angebracht ist, muß die zuständige Person sowohl über ein sehr detailliertes Wissen über die Probleme des laufenden Projekts, als auch über eine genaue Kenntnis der Inhalte der Experience Base verfügen.

Um diese Abhängigkeit von einer Einzelperson zu reduzieren, muß zumindest eine Möglichkeit geschaffen werden, mit der die Mitarbeiter der Projektorganisation selbst nach Erfahrungen in der Experience Base suchen können, die zur Lösung eines aktuellen Problems beitragen könnten. Als erster Schritt dazu wird in *VP-Admin* mit prototypischen Werkzeugen experimentiert, die elementare Suchmechanismen zumindest für die maschinell lesbaren Inhalte der Experience Base zur Verfügung stellen. Für eine darüber hinaus gehende Unterstützung ist eine Beschreibung der Zusammenhänge zwischen den einzelnen Informationen mit definierter Semantik erforderlich, woraus sich weitere Anforderungen an den in Abschnitt 3.1 skizzierten Beschreibungsformalismus ergeben.

4 Inhaltliche Ausrichtung der Experience Factory in *VP-Admin*

Wie eingangs erwähnt, ist die Experience Factory in *VP-Admin* in einem Bereich angesiedelt, der vorrangig mit der Entwicklung administrativer Softwaresysteme befaßt ist. Gerade in diesem Umfeld wird die Realisierung in zunehmenden Maß nicht selbst durchgeführt, sondern zu großen Teilen an externe Auftragnehmer vergeben. Damit erhält die Frage zusätzliches Gewicht, wie sichergestellt werden kann, daß Resultate, die von den Auftragnehmern geliefert werden, tatsächlich die erforderliche Qualität besitzen. Das noch vor einigen Jahren weit verbreitete Verständnis, daß der Entwicklungsprozeß beim Auftragnehmer für den Auftraggeber nicht transparent zu sein braucht und sich Qualitätssicherungsmaßnahmen ausschließlich auf die Produkte beziehen, hat sich mittlerweile als nicht ausreichend erwiesen.

Im Umfeld, in dem die Experience Factory in *VP-Admin* angesiedelt ist, wird versucht, bezüglich der Qualitätssicherung das Augenmerk auch verstärkt auf den Entwicklungsprozeß beim Auftragnehmer zu richten, z.B. um sicherzustellen, daß dort intern bereits Qualitätssicherungsmaßnahmen ergriffen werden, über die geeignete Nachweise geführt werden können. Qualitätssicherung auf Auftraggeberseite kann sich dann zum Teil darauf beschränken, die entsprechenden Qualitätsnachweise zu prüfen und ggf. noch durch eigene Stichprobenprüfungen zu ergänzen - eine Vorgehensweise, die etwa in der Automobilbranche im Verhältnis zu Zulieferern seit langem gängige Praxis ist und die sich z.B. auch in den Richtlinien zum Auftragnehmermanagement der ISO 9000 ff. widerspiegelt [Wal95, KeJ96].

Ein solcher Blickwinkel auf das Qualitätsmanagement hat weitreichende Auswirkungen. So müssen z.B. bereits bei der Vertragsgestaltung entsprechende Vorkehrungen getroffen werden. Aus den Erfahrungen, die im Umfeld dieses Verständnisses von Qualitätsmanagement gemacht werden, wird bislang noch nicht systematisch gelernt, d.h. es war in der Regel dem Zufall überlassen, daß von den gemachten Erfahrungen auch Personen profitieren konnten, die sie nicht selbst gemacht hatten.

Hier setzt die Experience Factory in *VP-Admin* an. *VP-Admin* stützt sich derzeit auf drei Entwicklungsprojekte, aus denen die Informationen und Erfahrungen für die Experience Base gezogen werden. Das Augenmerk liegt derzeit auf dem Thema Abnahmeprozesse. Die Informationen, die in diesem Zusammenhang momentan gesammelt und aufbereitet werden, sind neben der Charakterisierung des Entwicklungsprojekts, aus dem sie stammen, u.a.

- Beschreibung der Ergebnistypen wie etwa Spezifikations- oder Dokumentationsunterlagen, Quellcode, Qualitätsnachweise usw., die im Rahmen der Abnahme geprüft werden,
- Beschreibung der Abnahmeverfahren für die einzelnen Ergebnistypen,
- Beschreibung der zugrundegelegten Abnahmekriterien,

- Charakterisierung von Metriken zur Messung bestimmter Qualitätseigenschaften von Ergebnistypen,
- Angefallene Aufwände zur Durchführung von Abnahmeaktivitäten,
- Subjektive Aussagen, z.B. ob rückblickend Abnahmemaßnahmen den erwünschten Erfolg erbracht haben oder wo zusätzliche Maßnahmen erforderlich gewesen wären.

Darüber hinaus ist in VP-Admin geplant, gezielt Erfahrungen zu Reviews und Inspektionen als spezielles Abnahmeverfahren zu sammeln und zu einem späteren Zeitpunkt den Themenkreis Gestaltung von Spezifikationsdokumenten / Spezifikationsfähigkeit anzugehen.

Der Wert all dieser Erfahrungen liegt vor allem darin, daß sie hausinterne Vorgehensstandards und Richtlinien wie die ISO 9000 ff. dahingehend ergänzen, wie die gemachten Vorgaben tatsächlich im konkreten Projekt umgesetzt werden können. Zudem wird die Projektorganisation beim Versuch, vorhandene Erfahrungen zu nutzen, nicht sich selbst überlassen, sondern auch durch die Projektunterstützung der Experience Factory beraten.

5 Zusammenfassung

Das Projekt *VP-Admin* zielt zum einen darauf ab, mit Hilfe einer Experience Factory die Grundlage dafür zu legen, daß die Vergabe von Teilaufgaben in Softwareentwicklungsprojekten an externe Auftragnehmer wirksamer und effizienter abgewickelt und die Qualität der Ergebnisse zuverlässig geprüft werden kann. Dazu werden entsprechende Erfahrungen aus drei Entwicklungsprojekten systematisch gesammelt, aufbereitet und anderen Projekten wieder zur Verfügung gestellt. Der tatsächliche Nutzen einer Experience Factory für das Unternehmen kann erst im weiteren Verlauf von *VP-Admin* beurteilt werden, wenn die Erfahrungen tatsächlich von Entwicklungsprojekten genutzt werden.

Darüber hinaus werden in *VP-Admin* auch Erfahrungen auf einer Metaebene gesammelt, nämlich Erfahrungen, wie bei der Etablierung einer Experience Factory vorgegangen werden sollte und welche Probleme kognitiver, organisatorischer oder technischer Art sich dabei ergeben. Zusammen mit den Erfahrungen, die in zwei weiteren Projekten mit ähnlicher Zielsetzung, aber unterschiedlicher Anwendungsdomäne und Vorgehensweise bei der Einrichtung einer Experience Factory gemacht werden, kann so ein Beitrag dazu geleistet werden, das systematische Lernen aus Erfahrungen auch in anderen Bereichen des Unternehmens künftig effizient unterstützen zu können.

Zudem müssen in VP-Admin noch Fragen methodischer Art geklärt werden. Dazu zählen vor allem die Fragen, wie das Einbringen von Informationen in die Experience Factory und das Auffinden relevanter Erfahrungen etwa durch Werkzeuge unterstützt werden kann und wie die Inhalte der Experience Base geeignet strukturiert werden können.

Literatur

[Bas85] V.R. Basili: Quantitative evaluation of software engineering methodology. In *Proc. 1st Pan Pacific Computer Conference* (Melbourne, Australien), 1985

[Bas93] V.R. Basili: The experience factory and its relationship to other improvement paradigms. In *Software Engineering - ESEC'93*, I. Sommerville und M. Paul (Hrsg.), LNCS 717, Springer, Heidelberg, 1993, 68-83

[BCM92] V. Basili, G. Caldiera, F. McGarry, R. Pajerski, G. Page und S. Waligora: The Software Engineering Laboratory - An operational software experience factory. In *Proc. 14th International Conference on Software Engineering ICSE'92* (Melbourne, Australien), 1992, 370-381

[BCR94a] V.R. Basili, G. Caldiera und H.D. Rombach: Goal question metric paradigm. In [Mar94], 528-532

[BCR94b] V.R. Basili, G. Caldiera und H.D. Rombach: The experience factory. In [Mar94], 469-476

[GHJ94] E. Gamma, R. Helm, R. Johnson und J. Vlissides: *Design Patterns*. Addison-Wesley, Reading, 1994

[HoK97] F. Houdek und H. Kempter: Quality patterns - An approach to packaging software engineering experience. In *Proc. ACM Symposium on Software Reusability SSR'97* (Boston, Massachusetts), 1997

[KeJ86] R. Kehoe und A. Jarvis: *ISO 9000-3*. Springer, New York, 1996

[KuB94] P. Kuvaja und A. Bicego: Bootstrap - A European assessment methodology. In *Software Quality Journal 3(3)*, 1994, 117-127

[Mar94] J.J. Marciniak (Hrsg.): *Encyclopedia of Software Engineering, Band I*. Wiley, New York, 1994

[Min87] B. Minto: *The Pyramid Principle - Logic in Writing and Thinking*. Minto International, London, 3. Auflage, 1987

[MoC96] T.P. Moran und J.M. Carroll, Hrsg.: *Design Rationale - Concepts, Techniques, and Use*. Erlbaum, Mahwah, NJ, 1996

[PCC93] M.C. Paulk, B. Curtis, M.B. Chrissis und C.V. Weber: Capability Maturity Model, Version 1.1. In *IEEE Software 10(4)*, 1993, 18-27

[PWC95] M.C. Paulk, C.V. Weber, B. Curtis und M.B. Chrissis: *The Capability Maturity Model*. SEI Series in Software Engineering, Addison-Wesley, Reading, 1995.

[Rou95] T. Rout: SPICE: A framework for software process assessments. In *Software Process Improvement and Practice 1(1)*, 1995, 57-66

[Sen90] P. Senge: *The Fifth Discipline: The Art and Practice of the Learning Organisation*. New York, 1990

[Wal95] E. Wallmüller: *Ganzheitliches Qualitätsmanagement in der Informationsverarbeitung*. Hanser, München, 1995

Planung und Realisierung einer DV-Strategie

Axel Lukassen

Abstract

DV-Strategien werden oft in speziellen Projekten erarbeitet. In exklusiven Zirkeln werden High-Tech-Szenarien untersucht und unrealistische Maßnahmenpläne verabschiedet. Der hier vorgestellte Ansatz geht einen anderen Weg. In einem Arbeitskreis werden alle beteiligten Mitarbeiter zusammengeführt und Ergebnisse gemeinsam erarbeitet und diskutiert. DV-Strategie manifestiert sich bei diesem Ansatz nicht in alles umfassenden Gesamtansätzen sondern in kleinen realistischen Einzelschritten. Dem High-Tech-Diktat wird die Kraft von Teamwork und Langfristperspektive entgegengesetzt.

1 Einführung

Häufig ist es so, daß unklare Ziele und fehlende Strategien zu defensivem Vorgehen, zu Insellösungen und Zeitverzug und damit letztendlich zu Ineffektivität und Ineffizienz führen. In [Münz93] wurde dieser Regelkreis einmal als Negativspirale visualisiert. Ist dies einmal erkannt, so wird - oft unter Zeitdruck - ein Strategiefindungsprozeß gestartet. Hier sind unterschiedliche Szenarien denkbar. Einerseits besteht die Gefahr, daß realitätsfremde, idealisierte Ansätze entwickelt werden, die später nicht umgesetzt werden können. Andererseits kann es zu "Schnellschüssen" kommen, die unter Umständen mehr schaden als nützen. Ausgehend von einer nüchternen Bestandsaufnahme müssen Schwachstellen erkannt und Ziele formuliert werden. Die Marktsituation und mögliche Entwicklungsszenarien sind zu untersuchen.

Informatik ist ein Mittel zur Erreichung unternehmerischer Zielsetzungen. Die Wettbewerbsfähigkeit von Unternehmen hängt heute entscheidend von der Fähigkeit zur schnellen Analyse und Verarbeitung von Informationen ab. Daher kann auch die Entwicklung einer DV-Strategie nicht mehr nur technologieorientiert betrachtet werden. Zentrale Bestimmungsfaktoren sind aus der Businesstrategie herzuleiten.

In zunehmendem Maße wird erkannt, daß neben Businesstrategie und technologischen Faktoren der "Faktor Mensch" im DV-Management eine zentrale Rolle spielt. Anforderungen an das Management haben sich seit Beginn der Industrialisierung nachvollziehbar geändert. Waren in der Vergangenheit Manager gefragt, die technische Expertise nachweisen konnten, so sind heute Manager erfolgreich, die in der Lage sind, Mitarbeiter zu führen. Für den Strategieentwicklungsprozeß ist dies von großer Bedeutung. Die Rolle des DV-Managers kann es hier nicht mehr sein, der beste technische Experte zu sein, sondern seine Aufgabe besteht darin, die Potentiale seiner Mitarbeiter bestmöglich zur Entfaltung zu bringen. Nur wenn die Strategie nicht "von oben" vorgegeben wird, sondern im Team erarbeitet wird, besteht die Chance, daß Mitarbeiter sich die unternehmerischen Ziele zu eigen machen.

2 Grundlagen

Nach [Schr96] zielt die strategische Planung darauf ab, "den Bestand der Unternehmung dauerhaft sicherzustellen, d.h. es wird geprüft, ob in den derzeitigen Geschäftsfeldern mit dem jetzt gewählten Wettbewerbskonzept auch in Zukunft erfolgreich konkurriert werden kann".

Bei allen Beteiligten, aber zunächst einmal beim Top-Management des Unternehmens, ist die Bereitschaft gefordert, Dinge in Frage zu stellen, Diskussion zuzulassen, und Ergebnisse soweit zu konkretisieren, daß sie operationalisierbar werden. In den einzelnen Funktionsbereichen des Unternehmens sind nun Einzelstrategien zu entwickeln, die die Gesamtzielsetzung unterstützen. Dies gilt beispielsweise für Forschung und Entwicklung, Produktion, Service und für den Informatikbereich.

Gehen wir im folgenden davon aus, daß Informatik nicht das zentrale Geschäftsfeld eines Unternehmens ist. Dann stellen sich bei der Formulierung der Informatikstrategie beispielsweise folgende Fragen:

- Wie entwickelt sich der DV-Markt?
- Welche Anbieter werden den Markt dominieren?
- Welche Evolutions- bzw. Migrationsprozesse sind notwendig und denkbar?
- Welche Technologien werden sich durchsetzen?
- Wie lange sind alte Technologien noch zu halten?
- Wie sind Mitarbeiter auszubilden?
- Wie sehen die Rahmenbedingungen aus?
- Welche Produktlinie ist zu verfolgen (z.B. Win/OS2)?
- Ist Intranet ein Thema?
- Kann Informatik Wettbewerbsvorteile verschaffen?
- Kann Informatik Prozesse beschleunigen?
- Kann Informatik Qualität erhöhen?
- Sind durch Informatik kürzere Time-to-Market-Zyklen möglich?
- Wie gehen Konkurrenzunternehmen vor?

Informatikstrategie wird in [Lehner93] folgendermaßen charakterisiert: "Alle Maßnahmen zur Gestaltung der Informationsinfrastruktur sind darauf auszurichten, das in der Informationsfunktion vorhandene Erfolgspotential bestmöglich zur Wirkung zu bringen und in Unternehmenserfolg umzusetzen. Die Art und Richtung in der dies erfolgen soll, wird durch die Informatikstrategie beschrieben".

Die **Grundfrage der DV-Strategie** ist also: "Was ist zu tun, damit die Datenverarbeitung langfristig die entscheidenden Geschäftsprozesse so unterstützen kann, daß die Wertschöpfung in den Prozessen maximiert werden kann oder Informatik Wettbewerbsvorteile am Markt sichert oder verschafft ?"

Aufgrund der Dynamik des DV-Marktes ist einleuchtend, daß Informatikstrategien einem ständigen Anpassungsprozeß unterworfen werden müssen, wenn sie erfolgreich sein sollen. Bei der Implementierung des DV-Strategieprozesses kommt es daher ganz wesentlich darauf an, den Anpassungsprozeß mitzuimplementieren.

3 Ausgangslage, der Arbeitskreis

Die beschriebenen Erfahrungen beruhen auf der Gründung eines Arbeitskreises, der mit den Mitarbeitern des mittleren DV-Managements besetzt ist. Der Arbeitskreisansatz wurde gewählt, um den revolvierenden Charakter der Strategieentwicklung und -anpassung direkt mitzuimplementieren. Außerdem erscheint es notwendig, die Beteiligten nicht in einer Laborsituation aus dem normalen Arbeitsalltag herauszulösen. Darüber hinaus ermöglicht der Arbeitskreisansatz problemlos eine dynamische Erweiterung und Veränderung der personellen Zusammensetzung. Der Arbeitskreis erhielt den Titel DV-Architektur. Die Zielsetzungen des Arbeitskreises sind:

- Beschreibung eines Sollmodells der DV-Architektur
- Beschreibung eines langfristiges Szenarios
- Überprüfung bestehender Maßnahmenpläne
- Anstoß notwendiger Maßnahmen

Es ist eigentlich unerheblich, ob der Titel des Arbeitskreises den Begriff "DV-Strategie" enthält. Wesentlich ist nur, daß alle Beteiligten den konkreten Bezug der Arbeitskreisaktivitäten und der Zielsetzungen zum eigenen Umfeld erkennen können.

Der Gesamtprozeß beinhaltet 5 zyklisch wiederkehrende Phasen:

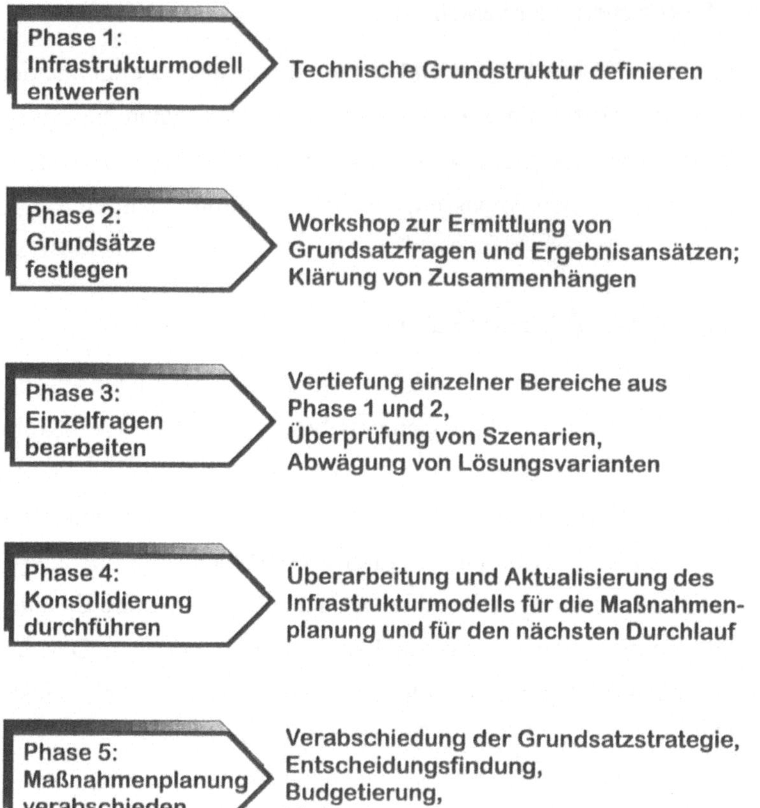

Der Arbeitskreis trifft sich im regelmäßigen Turnus zu Jour-Fix-Terminen (z.B. 14-tägig). Bei diesen Treffen werden Zwischenergebnisse diskutiert, Vereinbarungen zur weiteren Vorgehensweise getroffen, die Ausarbeitung einzelner Fragestellungen an einen oder mehrere Personen des Arbeitskreises delegiert. Je nach Bedarf können diese regelmäßigen Termine durch Workshops ergänzt werden. Durch gezielte Öffentlichkeitsarbeit im Unternehmen wird eine breite Akzeptanz für die erarbeiteten Ergebnisse geschaffen. Dieser Wechsel zwischen nach "Innen" und nach "Außen" gerichteten Aktivitäten des Arbeitskreises ist wesentlich für die Qualität der Ergebnisse.

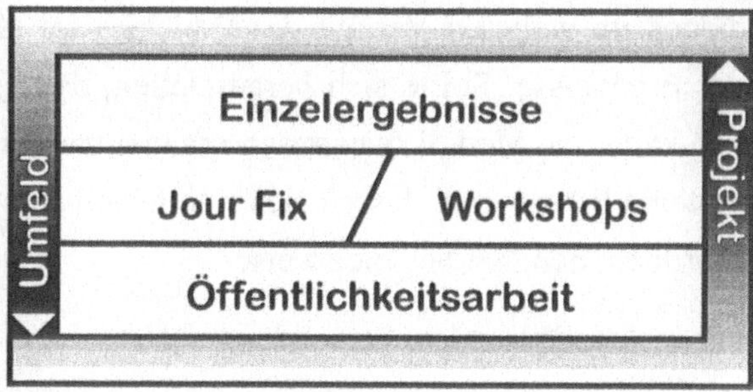

4 Das Infrastrukturmodell

Um eine Grundstruktur zur Orientierung zu haben, wurde zunächst ein DV-Infrastrukturmodell definiert. Dieses Infrastrukturmodell besteht aus den 4 Bereichen Anwendung, System, Hardware und Querschnittsfunktionen.

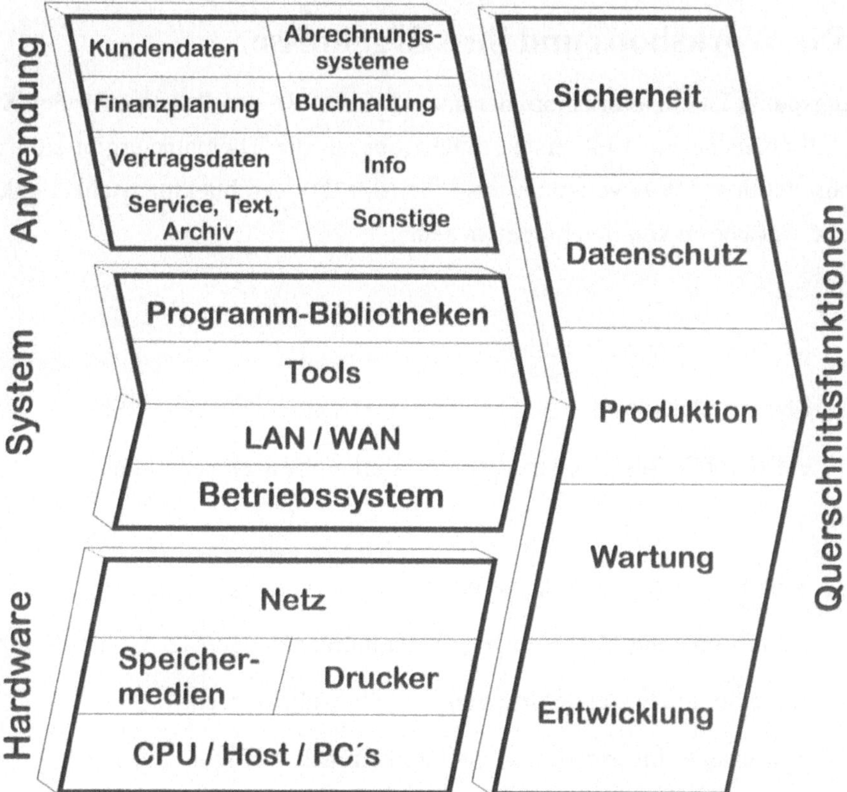

Diese Kategorien dienen im weiteren Verlauf des Projektes als Grundstruktur zur Einordnung einzelner Ergebnisse. Sollte sich herausstellen, daß einzelne Bereiche unvollständig sind, so kann das Modell dynamisch ergänzt werden. Die Aussagen, die die Informatikstrategie liefern muß, lassen sich nun jeweils einer oder mehrerer dieser Kategorien zuordnen, nämlich beispielsweise:

- Wo sind Schwachstellen in den Anwendungen?
- Wie wird sich die Anwendungslandschaft weiterentwickeln?
- Welche Anwendungsgebiete wird/muß es geben?

Der Fokus des Arbeitskreises kann je nach Bedarf ausgerichtet werden, z.B. auf Infrastruktur (HW, Automatisierung, Entwicklungsplattform), AE-Umgebung oder einzelne Querschnittsprobleme.

5 Die Workshops und ihre Ergebnisse

Ausgangspunkt ist ein Workshop, der dazu dienen soll, den Rahmen für den Strategieprozeß abzustecken. Eine erste Annäherung an die Thematik ergibt sich durch die Fragestellung: "Was veranlaßt uns, über die DV-Architektur nachzudenken?". Typische Antworten könnten hier etwa sein:

a) Probleme

- übergeordnete Ziele fehlen
- Mehrfrontenkampf
- existierende Systeme sind veraltet, starke physische Bindung

b) Ziele

- Verbesserung der AE-Umgebung
- Schaffung einer transparenten Systemumgebung
- Handeln, nicht von der Entwicklung überrollt werden
- gewünschte Information schnell und einfach verfügbar machen

- Wettbewerbsfähigkeit sichern

c) Menschen

- Ziele schaffen, mit denen man sich identifizieren kann
- Vorstellung von der Zukunft erarbeiten / vermitteln
- Wunsch nach gemeinsamer Zieldefinition

Ohne es näher untersucht zu haben, liegt die Vermutung nahe, daß die obigen Antworten so oder ähnlich in vielen Unternehmen das Ergebnis der Diskussion wären. Nachdem die generelle Motivation für den Strategieprozeß von allen Beteiligten formuliert wurde, geht es nun darum einzelne Ansätze für die Strategieentwicklung zu finden und zu formulieren. Die Erfahrung zeigt, daß dies für alle Beteiligten dann am leichtesten ist, wenn das eigene Umfeld in den Blickpunkt gerückt wird.

So kann die nächste Fragerunde unter der Überschrift: "Wie stellt sich ein Mitarbeiter der DV-Abteilung, bzw. der Fachabteilung seinen Arbeitsplatz in 5 Jahren vor?", gestartet werden. Aus Sicht der Fachabteilung ergeben sich beispielsweise folgende Antworten:

- Geschäftsvorfallorientierung
- Elektronische Akte
- Zeitnaher ad-hoc-Service
- Benutzerfreundliche Anwendungen, fehlertolerante Systeme
- Integration der Anwendungen
- Tests neuer Anwendungen automatisieren
- Parallele Sessions auf unterschiedlichen Rechnern
- 24-Std DV-Service ermöglichen

Natürlich wird deutlich, daß die Mitarbeiter der Fachabteilungen Wert auf Serviceorientierung legen und die Rolle der DV als Dienstleister im Unternehmen sehen.

Die gleiche Frage aus Sicht der DV kann beispielsweise zu folgenden Antworten führen:

- Transparente PC/Host-Welt ist sichergestellt
- Netzmanagement automatisiert
- Produktion automatisiert
- Geschäftsvorfallorientierung der Anwendungen
- Komfort der Anwendungsentwicklungs-Umgebung ist groß, z.B. Werkzeuge zum Prototyping, wiederverwendbare SW-Bausteine, abgestimmter Tooleinsatz, OO-Methodik
- Anwendungen sind benutzerfreundlich, z.B. Bedienung, Fehlertoleranz, graphische Oberfläche, auf Anwender abgestimmte Systemleistung (Anfängermodus, Expertenmodus)
- Electronic Mail zur asynchronen Kommunikation im ganzen Unternehmen

Die hier gefundenen Antworten dienen nun dazu, Einzelfragen herauszuarbeiten und zu priorisieren. Beispielsweise kann man hier die Bedeutung für Hauptprozesse des Unternehmens (wie Auftrag/Lieferung, Time-to-Market und Customer-Services) als Basis für die Beurteilung heranziehen und durch entsprechende Portfolios visualisieren.

6 Einzelfragen, Konsolidierung, Maßnahmenplanung

Die Ergebnisse der Workshops führen erfahrungsgemäß zunächst auf Problemstellungen, die die technische Infrastruktur betreffen, wie etwa Fragen des Systemmanagement oder der Anwendungsarchitektur. Entsprechend kann der Fokus daher erst einmal auf Infrastruktursachverhalte gelegt werden. Entsprechende Maßnahmen werden in die Planung eingebracht und umgesetzt.

Im Laufe der Zeit zeigen sich durch veränderte Rahmenbedingungen neue Fragestellungen und weiterer Handlungsbedarf. So können in späteren Arbeitsphasen des

Arbeitskreises andere Themen, wie etwa die Anwendungsentwicklungsarchitektur, im Mittelpunkt stehen.

Der weitere Verlauf in dieser Phase ist klassisch. Es werden Lösungsalternativen betrachtet, Entwicklungsszenarien beleuchtet, Machbarkeitsüberlegungen angestellt und Kostenentwicklungen untersucht. Auch hier werden Zwischenergebnisse im Arbeitskreis-Plenum vorgestellt, diskutiert und überprüft. Nach dem ersten Durchlauf werden die bisher erarbeiteten Ergebnisse in das Infrastrukturmodell eingearbeitet. Der Zusammenhang zu bestehenden Maßnahmenplänen wird kritisch durchleuchtet. Einzelfragen werden abschließend diskutiert, vertagt oder nochmals überprüft. Kostenbetrachtungen werden validiert. Wo notwendig werden Entscheidungsprozesse in Gang gesetzt und Maßnahmen in die turnusmäßige Planung eingebracht.

Gerade im Hinblick auf die Maßnahmenplanung ist es auch wichtig, daß die Ergebnisse des Arbeitskreises im Unternehmen publiziert und diskutiert werden. Hierdurch wird sichergestellt, daß der Arbeitskreis nicht in virtuelle Welten abdriftet.

7 Ergebnisse

Es wurde ein Verfahren zur Entwicklung einer DV-Strategie vorgestellt. Basis dieser Vorgehensweise ist die Gründung eines Arbeitskreises. Da die vorliegenden Erfahrungen auf der Arbeit in mittelständischen Unternehmen beruhen, stellt sich die Frage, wie sich ein derartiger Prozeß beispielsweise in einem Großkonzern initiieren läßt.

Die beschrieben Vorgehensweise besitzt folgende positiven Eigenschaften:

- Nicht der perfekte, einmalige Gesamtplan, sondern ein kontinuierlich ablaufender, langfristiger Prozeß wird implementiert
- Mitarbeiter gestalten den Strategieprozeß
- Viele bei Mitarbeitern vorhandene Ideen können aufgegriffen werden

- Mitarbeiter sind aktiv und entwickeln weiter
- Mitarbeiter haben einen nachvollziehbaren Entscheidungsrahmen
- Die Strategie kann dynamisch angepaßt und erweitert werden
- Die Fokussierung und Einschränkung auf bestimmte Themen ist möglich
- Mitarbeiter werden nicht in einer Laboratmosphäre aus der Arbeit herausgerissen
- Der Arbeitskreis kann dynamisch personell erweitert werden
- Wegen der langfristigen Perspektive kann aus Fehlern gelernt werden
- Zusammenarbeitsverhalten in der DV und zwischen den Abteilungen ändert sich
- Anspruch der DV-Strategie wird realistisch
- Ergebnisse sind pragmatisch und umsetzbar

Der beschriebene Ansatz führt zu handfesten Ergebnissen. So war es möglich innerhalb von 2 Jahren wesentliche Elemente der DV-Infrastruktur neu auszurichten und so die Grundlage für den Erfolg zentraler Anwendungsprojekte zu schaffen.

Oft ist es nicht eine fehlende Lösung, die den Erfolg verhindert, sondern mangelndes Zusammenarbeitsverhalten und fehlende Beteiligung der Betroffenen. Daher ist es ein zentrales Element dieser Vorgehensweise, regelmäßig im Arbeitskreisplenum zusammenzukommen und alle Betroffenen am Entwicklungsprozeß aktiv zu beteiligen. Nur eine Strategie, die von den Mitarbeitern getragen wird, kann erfolgreich sein.

Die Auswirkungen auf das Zusammenarbeitsverhalten sind nicht zu übersehen. Sowohl innerhalb der DV, als auch in der Kommunikation mit den Fachabteilungen ergibt sich ein neues gegenseitiges Verstehen und Problembewußtsein.

Literatur

[Schr96] G. Schreyögg: Strategische Planung, Fernuniversität Hagen 1988

[Münz93] H. Münzenberger: Strategische Informationsplanung, in Müller-Ettrich (Hrsg): Fachliche Modellierung von Informationssystemen, Addison-Wesley 1993

[Lehner93] F. Lehner: Informatikstrategien, Hanser 1993

Der Einsatz der Optionspreistheorie zur Bewertung von Softwareentwicklungsprojekten

Eberhard Stickel

Abstract

Die Beurteilung der Wirtschaftlichkeit von Softwareentwicklungsprojekten ist ein bedeutsames Problem, das bis jetzt aber leider nur rudimentär gelöst ist. Die Kapitalwertmethode wird in der Praxis oft eingesetzt. Sie erfordert detaillierte Zahlungsströme und basiert auf relativ restriktiven Prämissen, die bei praktischen Anwendungen nahezu nie erfüllt sind. Im vorliegenden Beitrag wird gezeigt, wie Ergebnisse der Optionspreistheorie angewendet werden können, um die Wirtschaftlichkeit von Softwareentwicklungsprojekten abzuschätzen. Der vorgestellte Ansatz erlaubt insbesondere die Berücksichtigung und Bewertung von Handlungsspielräumen der Entscheidungsträger. Die Methode wird anhand eines durchgängigen Beispiels, welches auf einem Praxisfall beruht, skizziert. Zum Abschluß werden mögliche Erweiterungen sowie noch zu lösende Probleme angegeben.

1 Einführung

Die Beurteilung der Wirtschaftlichkeit von Softwareentwicklungsprojekten ist ein in der Praxis äußerst bedeutsames Problem. Leider existieren bisher nur rudimentäre Ansätze, gelegentlich wird die Möglichkeit einer detaillierten Wirtschaftlichkeitsrechnung auf der Basis monetärer Größen gänzlich bezweifelt [Ant95].

Die Probleme liegen zum einen auf der ex ante Abschätzung von Entwicklungskosten (Problematik der Projektaufwandschätzung), zum anderen ist es sicherlich schwierig, beliebige Nutzeffekte zu monetarisieren. Dies gilt insbesondere vor dem Hintergrund moderner Softwaresysteme, deren Hauptnutzeffekte entgegen früheren Systemen nicht mehr primär im Bereich von direkt realisierbaren Kostensenkungspotentialen, sondern vielmehr im Bereich qualitativer Verbesserungen (Arbeitsqualität, Fehlerhäufigkeit, Verschiebung von Tätigkeitsstrukturen in Richtung höherwertiger Tätigkeiten) beziehungsweise im Bereich strategischer Nutzeffekte (verteidigbare Wettbewerbsvorteile, etwa Eintrittsbarrieren, Value Added Services o.ä.) liegen.

Die Betriebswirtschaftslehre stellt zahlreiche Investitionsrechenverfahren zur Verfügung. Großer Beliebtheit erfreut sich insbesondere die Kapitalwertmethode. Zu ihrer Anwendung werden detaillierte Ein- und Auszahlungsströme benötigt. Darüber hinaus ist die Angabe eines geeignete Diskontierungsfaktors erforderlich, um Zeitpräferenzen Rechnung zu tragen. Schon wenn es sich um sichere Zahlungsströme handelt, ist die Angabe dieses Diskontierungsfaktors ein in der Literatur intensiv diskutiertes Problem. Im allgemeinen sind die Zahlungsströme allerdings mit Unsicherheit behaftet. Dies erfordert die Verwendung risikoangepaßter Diskontierungsfaktoren. Letzteres ist ein in der Praxis nahezu unlösbares Problem.

Betrachtet man die Charakteristika von Softwareentwicklungsprojekten, so stellt man eine mehr oder weniger starke Gliederung aufgrund verwendeter Vorgehensmodelle fest. Typischerweise existieren mehrere sogenannte Meilensteine, bei deren Erreichen über die Fortführung des Projektes neu entschieden werden kann. Die Entscheidungsträger verfügen folglich über Handlungsalternativen, die man auch als Realoptionen interpretieren kann. Die Kapitalwertmethode kann derartige Handlungsspielräume nicht bzw. nur unter großen Schwierigkeiten bewerten.

In der Praxis beobachtet man immer wieder, daß Unternehmen im Rahmen von Pilotprojekten zunächst versuchen, Erfahrungen mit neuen Softwaretechnologien zu sammeln. Danach wird dann entschieden, ob es lohnend ist, die neue Technologie in größerem Umfang einzusetzen. Die Kapitalwertmethode vermag dieses Phänomen im allgemeinen nicht zu erklären. So ergibt sich regelmäßig ein negativer Kapitalwert. Pilotprojekte sollten demnach nie durchgeführt werden.

Im vorliegenden Beitrag wird gezeigt, wie die Optionspreistheorie zur Bewertung von Softwareentwicklungsprojekten herangezogen werden kann. Vorhandene

Handlungsspielräume werden dabei in natürlicher Weise in der Bewertung berücksichtigt. Die Methode kann sowohl ergänzend zur Kapitalwertmethode eingesetzt werden, diese aber auch substituieren. Dann werden weniger detaillierte Daten benötigt. Anschließende Sensitivitätsanalysen und gegebenenfalls weitere Simulationen sind in diesem Fall wesentlich einfacher möglich. Der Realoptionsansatz kann darüber hinaus Pilotprojekte rechtfertigen. Der Hauptgrund sind mögliche Optionen auf weitere Anwendungen, die mit einem Pilotprojekt "erworben" werden.

Die Anwendung des skizzierten Ansatzes hat Auswirkungen auf die Planung von Softwareentwicklungsprojekten. Es wird künftig ein bedeutender Erfolgsfaktor sein, Handlungsspielräume zu erkennen beziehungsweise derartige Handlungsspielräume gezielt zu schaffen. Ein Unternehmen, das in diesem Bereich erfolgreich ist, kann signifikante Wettbewerbsvorteile realisieren.

Der Beitrag ist wie folgt gegliedert. Im nächsten Abschnitt wird detailliert die Bedeutung der Wirtschaftlichkeitsanalyse für den Entwicklungsprozess von Software diskutiert. Es erfolgt eine Darstellung verfügbarer Ansätze sowie eine Kritik dieser Ansätze. In Abschnitt 3 werden die Defizite der Kapitalwertmethode erarbeitet. Daneben wird ein realitätsnahes Beispiel präsentiert, welches durchgehend innerhalb der Arbeit verwendet wird. Abschnitt 4 stellt den Realoptionsansatz ausführlich vor. Der Schwerpunkt liegt auf der Diskussion von Pilotprojekten, wie er in rudimentärer Form in [KHM95] vorgestellt wurde. Abschnitt 5 enthält einen Ausblick und faßt die erarbeiteten Ergebnisse kritisch zusammen.

2 Die Wirtschaftlichkeitsanalyse im Rahmen der Softwareentwicklung

Die Bedeutung einer detaillierten Wirtschaftlichkeitsanalyse im Vorfeld eines Softwareentwicklungsprojektes wird heute allgemein anerkannt. Unterschiedliche Auffassungen bestehen allenfalls bezüglich der Methodik beziehungsweise der Vorgehensweise. Betrachtet man ein Softwareentwicklungsprojekt als Investitionsprojekt, so steht ein ganzes Instrumentarium von Investitionsrechenverfahren zur

Verfügung. Dabei ist allerdings vorauszusetzen, daß es um die Bewertung eines einzelnen isolierbaren Projektes geht. Dies dürfte eine mit den Intentionen der Praxis grundsätzlich in Einklang stehende Annahme sein. Die integrierte Investitions- und Finanzplanung kann von den im folgenden betrachteten Rechenverfahren nicht geleistet werden kann. Eine detaillierte Behandlung dieses Problems kann und soll der vorliegende Beitrag nicht leisten. Zur Diskussion derartiger Fragen wird deshalb auf die Literatur verwiesen (vgl. etwa [Spr96, S. 431 ff.]).

Die isolierte Bewertung eines Projektes setzt zumindest die Separation der Investitionsentscheidung von den zugrunde liegenden Finanzierungsmaßnahmen voraus. Für eine ausführliche Diskussion der Bedingungen, unter denen eine derartige Separation möglich ist und die sich daraus ergebenden Konsequenzen sei erneut auf die Literatur verwiesen (vgl. [vNi97]).

Zur Anwendung von Investitionsrechenverfahren benötigt man im allgemeinen detaillierte Ein- und Auszahlungsströme während der Entwicklungs- beziehungsweise Nutzungszeit. Das bekannteste Verfahren, welches sehr häufig in der betrieblichen Praxis zur Bewertung von Investitionsprojekten eingesetzt wird, ist die Kapitalwertmethode. Ein- und Auszahlungsströme in einzelnen Perioden werden jeweils saldiert, anschließend erfolgt eine Diskontierung der saldierten Periodengrößen auf die Gegenwart und schließlich die Addition dieser diskontierten Zahlungsströme. Ein positiver Kapitalwert zeigt, daß das zugrunde liegende Investitionsprojekt einen Ertrag über der erwarteten Mindestrendite erzielt, ein negativer Kapitalwert impliziert genau das Gegenteil. Der gewählte Abzinsungssatz entspricht dabei in der Praxis der erwarteten Mindestrendite.

Neben der Kapitalwertmethode werden darüber hinaus die Interne Zinsfußmethode und die Amortisationsmethode eingesetzt. Schließlich werden in einigen wenigen Unternehmen immer noch Vergleichsrechnungen ohne Berücksichtigung der Zeitstruktur der anfallenden Ein- und Auszahlungen vorgenommen (Kosten-, Gewinn- bzw. Rentabilitätsvergleichsrechnungen). Man kann allerdings zeigen, daß nur die Kapitalwertmethode bestimmte durchaus plausible Mindestanforderungen erfüllen kann [CWe83; S. 26 ff.]. Dies sind im einzelnen die folgenden Forderungen:

- Berücksichtigung aller Ein- und Auszahlungen während der Laufzeit des Investitionsprojektes. Bezogen auf die Softwareentwicklung betrifft dies also den gesamten Systemlebenszyklus. Gegen diese Forderung verstößt beispielsweise das Verfahren der Amortisationsdauer.

- Berücksichtigung der Zeitstruktur der anfallenden Zahlungen. Demnach sind frühe Zahlungen anders zu gewichten als in der Höhe identische Zahlungen zu einem späteren Zeitpunkt. Gegen diese Forderung verstoßen insbesondere die auf Diskontierung der Zahlungsströme verzichtenden Vergleichsverfahren.

- Prinzip der Wertadditivität. Kann ein Projekt in mehrere voneinander unabhängige Teilprojekte zerlegt werden, so kann man jedes dieser Teilprojekte einer isolierten Bewertung unterziehen. Die Bewertung des Gesamtprojektes sollte sich dann aus der Summe der Bewertung der Einzelprojekte ergeben, so daß sich bezüglich der Beurteilung der Durchführung konsistente Entscheidungen ergeben. Gegen dieses Wertadditivitätsprinzip verstößt unter anderem die Methode des internen Zinsfußes.

Trotz der Tatsache, daß die Kapitalwertmethode diese Plausibilitätsforderungen erfüllt, wird ihre Anwendung insbesondere zur Beurteilung der Wirtschaftlichkeit von Softwareentwicklungsprojekten häufig stark kritisiert [Sch93], [Ant95]. Insbesondere wird in der Literatur betont, daß für Softwareprojekte detaillierte Zahlungsströme nur unter großen Schwierigkeiten beziehungsweise gar nicht zur Verfügung gestellt werden können [Sti92]. Es wird betont, daß dies besonders für die Einzahlungsströme, die ja die Nutzeffekte eines derartigen Anwendungssystems symbolisieren sollen, gelte. Zur Verdeutlichung wird dann ausgeführt, daß die häufigsten Nutzeffekte, die man mit heutigen Softwareentwicklungsprojekten realisieren will, qualitativer Natur sind. Dabei handelt es sich oftmals um höhere Vorgangsqualität (niedrigere Fehlerrate, höhere Kundenzufriedenheit usw.), sowie um Verschiebungen von Tätigkeitsstrukturen (vgl. etwa die Diskussion des hedonistischen Modells in [Sti92]).

Neben diesen qualitativen Effekten gewinnen in jüngerer Zeit verstärkt auch strategische Nutzeffekte an Bedeutung. Mit der Entwicklung wettbewerbsorientierter Systeme versuchen Unternehmen, mittel- bis langfristig verteidigbare Wettbewerbsvorteile gegenüber Mitbewerbern aufzubauen. Auch hier ist es schwierig, explizite Zahlungsströme anzugeben.

Akzeptiert man diese Kritik, so steht man vor einem gravierenden Dilemma. Auf der einen Seite können Softwareentwicklungsprojekte nicht losgelöst von einer Wirtschaftlichkeitsbetrachtung in Angriff genommen werden, auf der anderen Seite sind die bekannten Methoden und Verfahren, Wirtschaftlichkeit zu „messen" aufgrund der falschen Maßgröße nicht anwendbar. In der Literatur werden zahlreiche Verfahren diskutiert, die auf eine exakte Messung der Kosten- und Nutzeffekte verzichten. In [Ant95] wird ein Ansatz zur Beurteilung der Wirtschaftlichkeit auf der Basis von Wirtschaftlichkeitsprofilen vorgestellt. Es entsteht ein multikriterieller Ansatz. Bei multikriteriellen Verfahren muß in aller Regel dann eine Zusammenfassung der Kriterien auf der Basis von Nutzwerten oder Scoring-Verfahren vorgenommen werden. Die Vergleichbarkeit erfolgt dann durch eine künstliche und oft fragwürdige Skalierung. [Ant95] verzichtet auf die Zusammenfassung der Kriterien. Es erfolgt vielmehr ein Vergleich der Ist-Ausprägungen mit Soll-Ausprägungen. Letztere sind Mindestanforderungen, die keinesfalls verletzt werden dürfen. Das vorgestellte Verfahren kann folglich höchstens eine Erfüllung von Mindestkriterien garantieren. Eine Auswahl aus mehreren Alternativen (Argumentenbilanz) ist aufgrund der üblicherweise vorhandenen Kriterienvielfalt nur unter Schwierigkeiten möglich. So sind beispielsweise Projekte nur auf Basis von Dominanzkriterien vergleichbar.

Sind nun derartige Verfahren eine Lösung des oben skizzierten Dilemmas? Gerade im Zeitalter von Shareholder Value Konzepten (vgl. [Spr96, S. 459-494]), im Rahmen deren Implementierung Unternehmen dazu übergehen, Mindestrenditen für einzelne Bereich zu fordern, muß die Antwort klar negativ sein. Die Wirtschaftlichkeit eines Entwicklungsprojektes muß sich zu einem späteren Zeitpunkt zwangsläufig in der Gewinn- und Verlustrechnung des Unternehmens niederschlagen. Ansonsten sollte schon mit Rücksicht auf die Interessen der Anteilseigner auf ein solches Projekt verzichtet werden. Als Konsequenz ergibt sich die Notwendigkeit, Zahlungsströme zumindest grob zu schätzen. Dann können Simulationsrechnungen unter Variation einzelner Komponenten dieser Zahlungsströme sowie unter Variation des Diskontierungszinssatzes durchgeführt werden. Insofern bleibt die Kapitalwertmethode also wichtige Grundlage einer Wirtschaftlichkeitsanalyse für Softwareentwicklungsprojekte. Ergänzend dazu sollten Verfahren zum Einsatz kommen, die eine monetäre Bewertung von Qualitätsmerkmalen beziehungsweise Qualitätseffekten, sowie von Verschiebungen von Tätigkeitsprofilen leisten (vgl. [Sti92]). Ähnliches gilt für strategische Nutzeffekte (vgl. [Sti95]). Darüber hinaus sollten dann noch bestehende Defizite der Kapitalwertmethode, auf die im folgenden eingegangen wird, weitestgehend beseitigt werden.

3 Defizite der Kapitalwertmethode

Es wird im folgenden davon ausgegangen, daß eine zumindest grobe Schätzung von Ein- und Auszahlungsströmen erfolgt ist. In diesem Falle ist prinzipiell die Anwendung der Kapitalwertmethode möglich.

Die Kapitalwertmethode geht davon aus, daß die zugrunde liegenden Zahlungsströme deterministische Größen sind. Wird die im allgemeinen sicherlich vorhandene Unsicherheit der Zahlungsströme berücksichtigt, sind risikoangepaßte Diskontierungsfaktoren in jeder Periode zu bestimmen. Diese risikoangepaßten Diskontierungsfaktoren variieren darüber hinaus noch in Abhängigkeit vom eingetretenen Umweltzustand. Konkret bedeutet dies, daß nun je nach Periode und Umweltzustand mit unterschiedlichen Diskontierungssätzen gerechnet werden muß. Die Bestimmung derartiger Zinssätze stellt bereits im Einperiodenfall ein nicht triviales Problem dar. Dort kann auf das Capital Asset Pricing Modell (CAPM) zurückgegriffen werden. Eine Übertragung dieses Modells auf den Mehrperiodenfall ist allerdings unmöglich [Lau93].

Ein einmal durch die Kapitalwertmethode als durchführungswürdig bezeichnetes Projekt wird nun als unteilbares ganzes begriffen und durchgeführt. Die Realität zeigt jedoch, daß während der Laufzeit eines Investitionsprojektes diverse Handlungsspielräume vorhanden sind, die ein Unternehmen zu seinem Vorteil ausnutzen kann. Zu denken wäre hier speziell im Falle von Softwareentwicklungsprojekten an die folgenden Möglichkeiten:

- Aufteilung eines Entwicklungsprojektes in Teilprojekte. Nach Durchführung der Teilprojekte kann jeweils über die Weiterführung oder über den Abbruch entscheiden werden. Zu diesen Zeitpunkten verfügt man über verbesserte Informationen, so daß der Grad der Unsicherheit der Ein- und Auszahlungen reduziert werden kann.
- Möglichkeiten zur Erhöhung des Projektumfangs bzw. der Kapazitäten im Falle besonders guter Markt- beziehungsweise Umweltkonstellationen.
- Möglichkeiten zur temporären Unterbrechung beziehungsweise zur zeitlichen Streckung eines Entwicklungsprojektes. Als Beispiel könnte die Durchführung eines Pilotprojektes dienen, um erste Erfahrungen mit einer neuen Softwaretechnologie zu sammeln. Gegebenenfalls kann man nach Ablauf des Pilotprojektes zunächst weitere Entwicklungen abwarten.

Alle diese Optionen können im Rahmen der klassischen Kapitalwertmethode nicht berücksichtigt werden. Grundsätzlich ist davon auszugehen, daß die Existenz derartiger Handlungsspielräume den Wert eines Softwareentwicklungsprojektes nachhaltig erhöhen kann. Folglich tendiert die Kapitalwertmethode in diesem Bereich zu einer sukzessiven Unterschätzung der Wirtschaftlichkeit. Zur Verdeutlichung der eben skizzierten Kritikpunkte soll das folgende einfache Beispiel betrachtet werden. Die Durchführung eines Softwareentwicklungsprojektes ist auf die nachfolgend skizzierten beiden Arten möglich.

1. Zu Beginn des Entwicklungszeitraumes (Periode 1) ist eine Anfangsinvestition in Höhe von g Geldeinheiten erforderlich. Ab der dritten Periode ergeben sich

für weitere fünf Jahre Rückflüsse in Höhe von jeweils r Geldeinheiten. Dabei sei r der Erwartungswert einer stochastischen Zufallsvariablen.

2. Zu Beginn des Entwicklungszeitraumes ist im ersten Jahr eine Anfangsinvestition in Höhe von $g/2$ Geldeinheiten erforderlich. Zu Beginn der zweiten Periode sind weitere $(1+i)g/2$ Geldeinheiten notwendig. Ab der dritten Periode ergeben sich Rückflüsse für die nächsten fünf Jahre in Höhe von jeweils r Geldeinheiten. Erneut sei r der Erwartungswert einer stochastischen Zufallsvariablen.

Unterstellt man einen Diskontierungszinssatz von i, so bewertet die Kapitalwertmethode beide Durchführungsvarianten gleich. Für die erste Variante ergibt sich nämlich der Kapitalwert

$$K_1 = -g + \sum_{j=2}^{6} r(1+i)^{-j},$$

für die zweite Variante erhält man entsprechend

$$K_2 = -\frac{g}{2} - (1+i)\frac{g}{2}(1+i)^{-1} + \sum_{j=2}^{6} r(1+i)^{-j} = K_1.$$

Die zweite Variante ist allerdings die wertvollere Variante, da sie mehr Gestaltungsmöglichkeiten zuläßt. So kann man das Projekt am Ende der ersten Periode abbrechen, wenn man feststellt, daß aufgrund eingetretener Umweltzustände die geplanten Rückflüsse nicht erreichbar sind. In diesem Fall kann auf den Zuschuß weiterer Mittel verzichtet werden. Zur Verdeutlichung sollen die folgenden konkreten Zahlenwerte dienen. Es sei g = 100.000 DM und i = 7%. Bei den Rückflüssen rechnet man im optimistischen Fall mit $r = r_o$ = 50.000 DM, im pessimistischen Fall dagegen mit $r = r_p$ = 10.000 DM. Beide Umweltzustände (optimistisch beziehungsweise pessimistisch) seien gleich wahrscheinlich und können am Ende der ersten Periode beobachtet werden (der Erwartungswert ist r = 30.000 DM).

Für die erste Variante ergibt sich im optimistischen Fall ein Kapitalwert von K_o = 91.598,01 DM, im pessimistischen Fall dagegen von K_p = -61.680,40 DM. Der Erwartungswert ergibt sich folglich zu K_1 = 14.958,81 DM. Wird die Entscheidung auf Basis dieser Daten getroffen, so wird das Projekt durchgeführt. In diesem Fall betragen die Entwicklungskosten DM 100.000, die Rückflüsse ergeben sich je nach Umweltzustand.

Bei der zweiten Projektvariante wird aufgrund des positiven Kapitalwerts mit der Durchführung des Projektes begonnen. Deshalb werden zunächst DM 50.000 investiert. Vor der zweiten Rate der Investition kann allerdings die bestehende Unsicherheit durch Beobachtung des eingetretenen Umweltzustandes eliminiert werden (in der Realität ist dies sicher nur teilweise der Fall, die Annahme dient lediglich der Vereinfachung der Darstellung). Tritt das optimistische Szenario ein, so wird ein Betrag in Höhe von DM 53.500 investiert. Danach ergeben sich Rückflüsse in Höhe von jährlich DM 50.000, so daß der Kapitalwert K_o realisiert wird. Tritt dagegen das pessimistische Szenario ein, so hat der Entscheidungsträger zwei Möglichkeiten. Er kann zum einen die weitere Investition tätigen, um durch die immer noch positiven Rückflüsse zumindest einen Teil der bereits aufgelaufenen Kosten zu kompensieren. Alternativ kann jedoch das Projekt jetzt auch abgebrochen werden. In unserem Fall ergibt sich bei Fortführung ein Kapitalwert von (r = 10.000 DM)

$$K_f = -\frac{g}{2}(1+i) + \sum_{j=1}^{5} r_p (1+i)^{-j} = -12.498{,}03 \text{ DM}.$$

Dabei ist zu beachten, daß jetzt der Beginn der zweiten Periode als Bezugsbasis dient. Der bereits investierte Betrag von DM 50.000 stellt sogenannte *sunk costs* dar und ist nicht mehr entscheidungsrelevant. Die Handlungsempfehlung lautet folglich, das Projekt abzubrechen, wenn der pessimistische Fall eintritt.

Der Unterschied zwischen beiden Projekten ist nun offensichtlich. Im zweiten Projekt besteht für das Unternehmen ein Handlungsspielraum. Es hat nämlich die Option, das Projekt ohne weitere Investitionen abzubrechen. In diesem Fall sind lediglich DM 50.000, statt eines sonst auftretenden Verlustes von DM 61.680,40 verloren. Der Wert der Handlungsoption beträgt folglich die Hälfte von DM 11.680,40 oder DM 5.840,20 (der Faktor 0,5 ist die Eintrittswahrscheinlichkeit des pessimistischen Szenarios, in welchem die Handlungsoption bedeutsam wird).

Als Fazit ist festzuhalten, daß die zweite Projektvariante einen höheren Wert für das Unternehmen besitzt, obwohl die Kapitalwertmethode beide Projektalternativen identisch bewertet. Dies verdeutlicht, wie wichtig das Erkennen und die daran anschließende Bewertung derartiger Handlungsalternativen sind. Dies gilt in besonderem Maße für die Softwareentwicklung, wo gestaffelte Investitionen in Form sogenannter Pilotprojekt sehr häufig auftreten. Das eben Gesagte gilt natürlich auch für Vorgehensmodelle mit Meilensteinen zum Ende einer Phase.

4 Realoptionen

Realoptionen sind Handlungsspielräume im Leistungsbereich eines Unternehmens, über die ein Entscheidungsträger verfügen kann. Insofern besteht eine enge Verbindung zu den bekannteren Finanzoptionen, die Rechte auf einen zugrunde liegenden Vermögensgegenstand verbriefen. Andererseits existieren aber auch gravierende Unterschiede. Während Finanzoptionen dem Entscheidungsträger exklusiv zur Verfügung stehen, ist dies bei Realoptionen nicht notwendigerweise der Fall. Realoptionen können auch den Mitbewerbern offen stehen. In diesem Fall sind zur Ermittlung des Wertes der Realoptionen die Aktionen potentieller Mitbewerber zu berücksichtigen. Als Beispiel sei die Option auf Errichtung eines wettbewerbsorientierten Informationssystems betrachtet. Hier sind die Reaktionen der Mitbewerber für den Erfolg eigener Aktionen oft maßgeblich verantwortlich [Sti95]. Bei Finanzoptionen wird der zugrunde liegende Vermögensgegenstand im allgemeinen an einer Börse gehandelt, so daß ein Marktpreis verfügbar ist. Auch dies ist bei Realoptionen so oft nicht der Fall.

Die Bewertung von Finanzoptionen erfolgt durch Arbitrageüberlegungen. Durch Kauf und Verkauf von Finanztiteln wird eine risikolose Anlage konstruiert, die sich zum risikofreien Zinssatz verzinsen muß. Eine analoge Vorgehensweise ist für Realoptionen möglich. Im folgenden soll der Bewertungsvorgang für eine Realoption unter Verwendung des Binomialmodells von Cox, Ross und Rubinstein skizziert werden [CRR79]. Zu diesem Zweck wird auf das Beispiel des letzten Abschnitts zurückgegriffen, ohne allerdings zuerst spezifische Werte für die Variablen anzunehmen.

Der Entscheidungsträger besitzt im Beispiel die Option, bei Vorliegen positiver Informationen das Investititionsprojekt nach einer Periode weiterzuführen (beziehungsweise ohne weitere Investitionen abzubrechen). Geht man von einem Wert der Erträge des Projektes in Höhe von S zu Beginn der ersten Periode aus, so ergibt sich im optimistischen Fall ein Wert von uS, im pessimistischen Fall dagegen ein Wert von dS. Die Realisation ist nach einer Periode bekannt. Der optimistische Fall trete mit Wahrscheinlichkeit q, der pessimistische Fall mit Wahrscheinlichkeit $1\text{-}q$ ein. Abb. 1 verdeutlicht diese Zusammenhänge nochmals in grafischer Form.

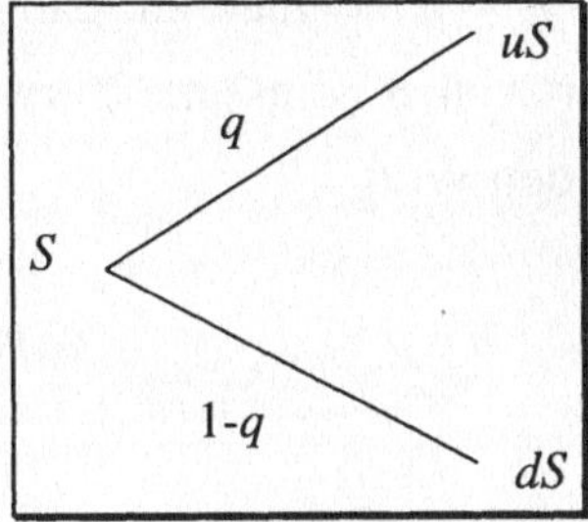

Abb. 1: Entwicklung des Investitionsprojektes im Zeitablauf

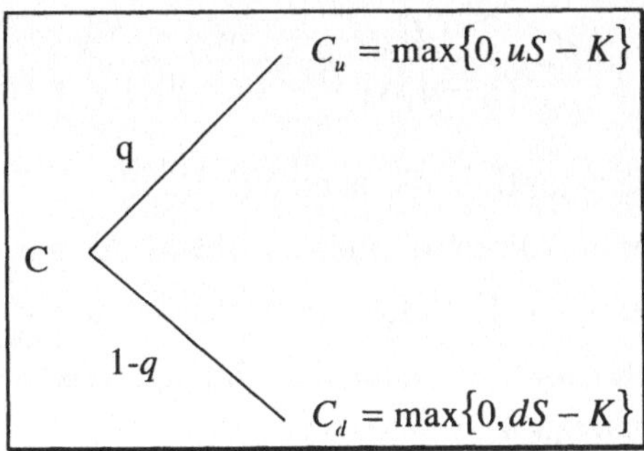

Abb. 2: Entwicklung des Optionswertes

Parallel dazu wird nun der Wert der Realoption betrachtet. Es handelt sich um eine Calloption auf das Projekt von Abb. 1. Der Wert der Option beträgt

$$C_u = \max\{0, uS - K\} \text{ bzw. } C_d = \max\{0, dS - K\}$$

im optimistischen beziehungsweise pessimistischen Fall (vgl. Abb. 2). Dabei ist K der sogenannte Basispreis, der im vorliegenden Fall mit der zweiten Rate $(1+i)g/2$ übereinstimmt. Der Basispreis ist zu entrichten, wenn die Option auf Weiterführung ausgeübt wird. Im nächsten Schritt soll nun ein risikofreies Portefeuille zum Zinssatz i konstruiert werden. Dazu nehmen wir fiktiv an, daß es ein an der Börse gehandeltes Wertpapier gibt, welches die Entwicklung unseres Investitionsprojektes exakt dupliziert. Darüber hinaus setzen wir Risikoneutralität des Entscheidungsträgers voraus. Der Wert des gehandelten Wertpapiers am heutigen Tage ist dann S, der Wert am Ende einer Periode beträgt uS mit Wahrscheinlichkeit q und dS mit Wahrscheinlichkeit $1-q$. Wir nehmen an, daß M Stück dieses Wertpapiers und risikofreie Bonds im Wert von B (Zinssatz i) erworben werden. Dieses Portefeuille hat nach einer Periode den Wert

$$C_u = MuS + (1+i)B \text{ bzw. } C_d = MdS + (1+i)B.$$

im optimistischen beziehungsweise pessimistischen Szenario. Die Auflösung nach M beziehungsweise B liefert sofort die Beziehungen

$$M = \frac{C_u - C_d}{(u-d)S},\ B = \frac{uC_d - dC_u}{(u-d)(1+i)}.$$

Verkauft man nun die Realoption, die ja eine Calloption auf das Investitionsprojekt darstellt zum (unbekannten) Preis C und erwirbt M Aktien sowie Bonds im Wert von B, so wurde ein risikofreies Portefeuille konstruiert. Folglich muß $C = MS + B$ gelten. Setzt man die oben errechneten Werte ein, so erhält man für den Optionswert die Gleichung.

$$C = \frac{1}{1+i}\left(\frac{1+i-d}{u-d}C_u + \frac{u-1-i}{u-d}C_d\right).$$

Dabei sollte sinnvollerweise $d<1+i<u$ vorausgesetzt werden. Sonst sollte das Projekt in keinem Fall durchgeführt werden ($1+i>u$ bedeutet, daß die Maximalverzinsung des Projektes immer noch unter der risikofreien Verzinsung liegt) beziehungsweise in jedem Fall durchgeführt werden ($1+i<d$ bedeutet, daß die Verzinsung des Projektes in jedem Fall über der risikofreien Rate liegt; bei Risikoneutralität impliziert dies die Durchführung unabhängig vom später eintretenden Szenario). Erwähnenswert ist, daß die Eintrittswahrscheinlichkeiten q beziehungsweise $1-q$ in der Optionspreisformel nicht mehr auftreten.

Im Beispiel des letzten Abschnittes erhält man nun die Zahlenwerte K = 53.500 DM und i = 7%. Weiter gilt

$$S = \frac{1}{1+i} E\left(\sum_{j=1}^{5} r(1+i)^{-j}\right) = \frac{1}{2(1+i)}(r_o + r_p)\sum_{j=1}^{5}(1+i)^{-j}.$$

Dabei bezeichnet E den Erwartungswert bezüglich der stochastischen Größe r. Im Beispiel folgt S = 114.958,81 DM. Für uS beziehungsweise dS ergibt sich

$$uS = r_o \sum_{j=1}^{5}(1+i)^{-j}, \; dS = r_p \sum_{j=1}^{5}(1+i)^{-j}.$$

also uS = 205.009,87 DM und dS = 41.001,97 DM. Daraus errechnet man die Parameter u = 1,783 und d = 0,357, sowie C_d = 0 DM und C_u = 151.509,87 DM. Daraus ergibt sich schließlich C = 70.799,01 DM. Die Calloption auf das Projekt hat folglich den Wert 70.799,01. Zieht man davon die Anfangsinvestition in Höhe von DM 50.000 und den erwarteten Kapitalwert K_1 des Projektes ab, so ergibt sich der Wert des Handlungsspielraumes "Projektabbruch" nach einer Periode zu DM 5.840,20. Dies stimmt mit den Überlegungen des vorigen Abschnitts überein. Im diskutierten Beispiel wurde der Optionsansatz ergänzend zur Kapitalwertmethode eingesetzt.

5 Zusammenfassung und Ausblick

Das vorgestellte Bewertungsmodell kennt im wesentlichen nur zwei Ausprägungen von Unsicherheit, nämlich die Zustände "Boom" und "Bust". Eine Verallgemeinerung auf mehr Zustände ist möglich (vgl. [Kul95]). Zur Ermittlung des Wertes der Calloption sind, falls auf die Kapitalwertmethode verzichtet wird, die folgenden Daten zu erheben beziehungsweise zu schätzen:

- erwarteter Ertrag uS (dS) im optimistischen (pessimistischen) Fall;
- erwartete Rendite u (d) im optimistischen (pessimistischen) Fall;
- Basispreis K der Calloption (im Beispiel zweite Rate der Investition);
- risikoloser Zinssatz i.

Da man im Regelfall den Wert der Anfangsinvestition kennt, kann der Wert des Handlungsspielraumes explizit errechnet werden. Im letzten Abschnitt wurde der Wert des Handlungsspielraumes auf Basis der Kapitalwertmethode errechnet. Mit den dort verwendeten Zahlungsströme können die Größen uS, dS, u, d und K geschätzt werden. Als einzigen Abzinsungsfaktor benötigt man den risikolosen Zinssatz i. Bei der Kapitalwertmethode sollte dagegen, anders als im Beispiel des letzten Abschnittes, mit risikoangepassten Abzinsungsfaktoren gerechnet werden. Darauf kann bei Realoptionen verzichtet werden. Dies stellt einen wesentlichen Vorteil dieses Ansatzes dar. Verzichtet man auf die Kapitalwertmethode, sind beim Optionsansatz nur die Größen uS beziehungsweise dS zu schätzen, so daß detaillierte Zahlungsreihen nicht notwendig sind. Gerade im Hinblick auf die Problematik der Ermittlung derartiger Zahlungsströme im Falle von Softwareentwicklungsprojekten ist dies ein weiterer Vorteil. Zur Bewertung von Investitionsprojekten sollten die Größen uS, dS, u sowie d im Sinne einer Sensitivitätsanalyse beziehungsweise durch Simulationen variiert werden, um die Auswirkungen auf den erwarteten Wert des Projektes zu bestimmen.

Gerade bei Softwareentwicklungsprojekten bestehen oft mehrere explizite Meilensteine, so daß sich eine Folge von Handlungsoptionen, die voneinander abhängen (Compund Options), ergibt. Dies und auch andere Optionsarten (z.B. Kapazitätsverschiebungen, temporäre Stillegung) können berücksichtigt werden. Abschließend sei erwähnt, daß sich das entwickelte Instrumentarium auch zur Bewertung von bestimmten Vorgehensmodellen eignet. Zu denken ist hier an eine Bewertung der einzelnen Zyklen innerhalb eines Prototypingansatzes.

Problematisch ist die Bestimmung der am Kapitalmarkt gehandelten Alternativinvestition. Hier kann auf Ergebnisse der Arbitrage Pricing Theorie zurückgegriffen werden [CWe83, S. 364 ff]. Insgesamt besteht hier allerdings weiterer Forschungsbedarf (vgl. auch die Diskussion in [Kul95] in bezug auf die Problematik von Renditeabschlägen bei nicht an Börsen gehandelten Anlagemöglichkeiten).

Literaturverzeichnis

[Ant95] Antweiler, J.: Wirtschaftlichkeitsanalyse von Informations- und Kommunikationssystemen auf der Basis von Wirtschaftlichkeitsprofilen. Information Management 10, Nr. 4, 1995, S. 56-64.

[CWe83] Copeland, T. E., Weston, J. F.: Financial Theory and Corporate Policy. 2. Aufl., Reading, MA. 1983.

[CRR79] Cox, J.; Ross, S., Rubinstein, M.: Option Pricing: A Simplified Approach. J. of Financial Economics 7, 1979, S. 229-263.

[KHM95] Kambil, A., Henderson, J., Mohsenzadeh, H.: Strategic Management of Information Technology Investments: An Option Perspective. In: Khsrowpour, M. (Ed.): Managing Information Technology Investments with Outsourcing. Harrisburg, PA, 1995, S. 32-54.

[Kul95] Kulatilaka, N.: The Value of Flexibility: A General Model of Real Options. Methods for Evaluating Capital Investment Decisions under Uncertainty. In: Trigeorgis, L.: Real Options in Capital Investment. Models, Strategies and Applications. Westport CT 1995, S. 89-107.

[Lau93] Laux, C.: Handlungsspielräume im Leistungsbereich des Unternehmens: Eine Anwendung der Optionspreistheorie. Zeitschrift für betriebswirtschaftliche Forschung (zfbf) 45, Nr. 11, 1993, S. 933-957.

[vNi97] Von Nitzsch, R.: Separation bei betrieblichen Investitionsentscheidungen. OR-Spektrum 19, Nr. 1, 1997, S. 55-65.

[Sch93] Schumann, M.: Wirtschaftlichkeitsbeurteilung für IV-Systeme. Wirtschaftsinformatik 35, Heft 2, 1993, S. 167-178.

[Spr96] Spremann, K.: Wirtschaft, Investition und Finanzierung. 5. Aufl., München, Wien (1996).

[Sti92] Stickel, E.: Eine Erweiterung des hedonistischen Verfahrens zur Ermittlung der Wirtschaftlichkeit des Einsatzes von Informationstechnik. Zeitschrift für Betriebswirtschaft (zfb) 62, Nr. 7, 1992, S. 47-64.

[Sti95] Stickel, E.: Wettbewerbsorientierte Informationssysteme und Produktivitätsparadoxon. Wirtschaftsinformatik 37, Heft 6, 1995, S. 548-557.

Erfolgsfaktoren des Managements von Migrationsprojekten

Fabian Dömer

Abstract

Eine signifikante Verbesserung der informationstechnischen Unterstützung bei gleichzeitiger Begrenzung des Aufwands ist in vielen Situationen nur über die Migration eines alten Informationssystems auf ein neues realisierbar. Die betroffenen Unternehmen sehen sich dabei häufig mit bisher unbekannten Aufgaben und komplexen Fragestellungen konfrontiert. Dieser Beitrag zeigt auf Basis einer empirischen Untersuchung von Migrationsprojekten, welche wesentlichen Faktoren den Erfolg einer Migration maßgeblich beeinflussen und bietet damit einen Orientierungsrahmen für die Planung und Durchführung von Migrationsprojekten.

1 Einleitung

Aufgrund des allgemein zunehmenden Durchdringungsgrades von Informationstechnologie (IT) in den Unternehmen vollzieht sich die Einführung von Informationssystemen (IS) immer seltener als Totalinnovation. Vielmehr müssen häufig bestehende IS im Rahmen einer Migration durch neue abgelöst werden, wobei die wertvollen Investitionen in Funktionen und Daten nach Möglichkeit zu schützen sind. Dabei ist keine Routinearbeit gefragt - Migrationsprojekte haben meist Pilotcharakter auf hohem Komplexitätsniveau. In der Konsequenz erreichen Migrationsprojekte häufig nicht das ursprünglich anvisierte Ziel oder überschreiten die geplante Projektdauer deutlich.

Als erfolgreich wird ein Migrationsprojekt eingestuft, wenn es sein Aufwands- und Terminsoll erfüllt, den Zielen der technischen Spezifikation entsprochen wird, die

Anwender qualifiziert, motiviert und zufrieden sind sowie die weiteren Wirtschaftlichkeitsziele (z.B. Erhöhung der Kundenzufriedenheit) erreicht sind.

Was macht Migrationen erfolgreich? Woran scheitern Migrationen? Was muß beim Management von Migrationsprojekten beachtet werden? Zur Klärung dieser Sachverhalte wurden im Rahmen einer empirischen Studie Migrationsprojekte unterschiedlicher Dimension und Inhalte analysiert. Die folgende Übersicht zeigt die ermittelten 30 Erfolgsfaktoren der Migration in Kurzform; sie kann als Checkliste für das Projektmanagement dienen.

2 Zum Begriff Migration

IS erreichen früher oder später ein Stadium, in dem sie den funktionalen oder technischen Anforderungen nicht mehr gerecht werden, gleichzeitig aber das Optimierungspotential durch Wartungsaktivitäten ausgereizt ist. Eine totale Neuentwicklung des Altsystems ist nur in den wenigsten Fällen durchführbar. Die Migration bietet hier mit einer Mischung aus Innovation und weitestgehender Wiederverwendung von Elementen des Altsystems die Möglichkeit, mit begrenztem Aufwand einen effektiven Technologiesprung zu erreichen. Die Migration ist begrifflich damit einerseits von der Wartung und andererseits von der Totalinnovation abzugrenzen (Abbildung 1).

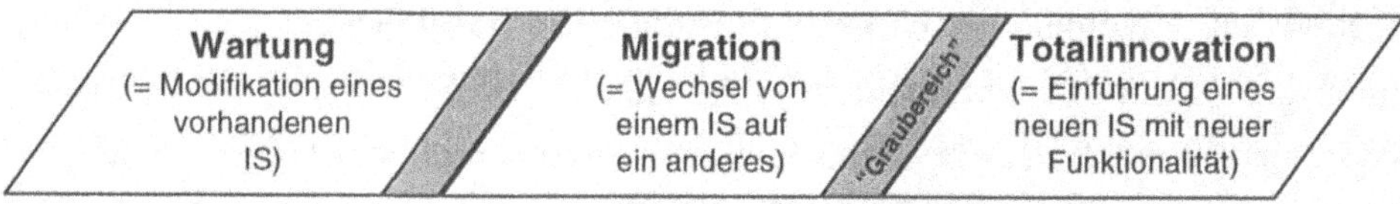

Abbildung 1: Migration in Abgrenzung zu Wartung und Totalinnovation

"Informationstechnik-Migration (IT-Migration) ist definiert als der Prozeß der Umstellung eines IS (= Altsystem/Quellsystem) auf ein anderes IS (= Neusystem/Zielsystem) im Rahmen einer eigenen Projektorganisation, wobei sich Alt- und Neusystem signifikant bzgl. der Infrastruktur i.e.S. (technische Komponenten: Hardware/Software) unterscheiden, gleichzeitig aber elementare *fachliche* und/oder *technische* Komponenten (Hardware/Software in beliebigen Repräsentationsformen und/oder Daten des Altsystems) übernommen werden. Ziel ist die Verbesserung der Leistungsfähigkeit, Qualität und/oder Wirtschaftlichkeit durch das Neusystem." [Döm97]

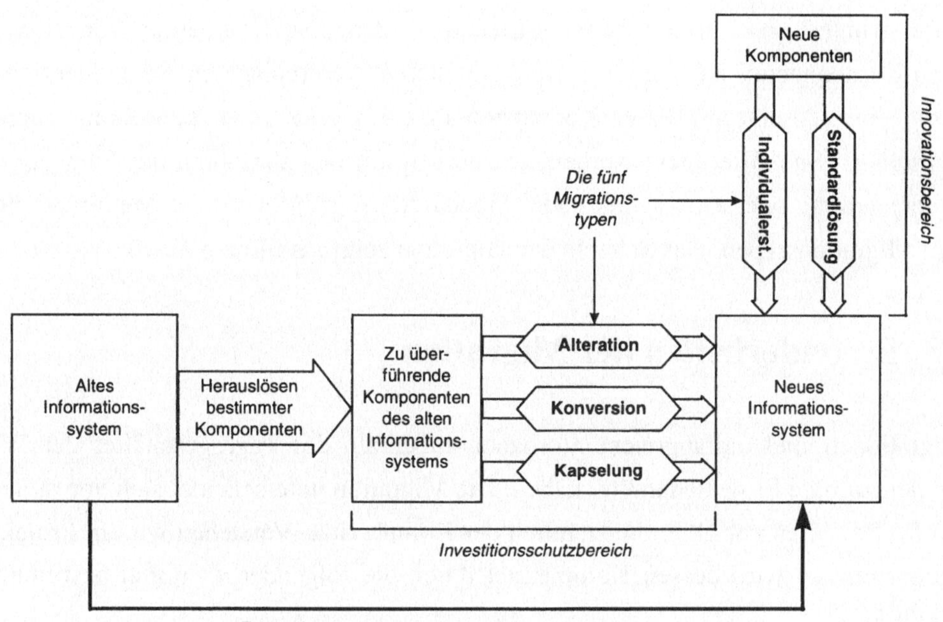

Abbildung 2: Migration - aus alt mach neu

Die technischen Bestandteile des IS definieren wir als Komponenten, z.B. die Benutzerschnittstelle, ein Anwendungsprogramm oder bestimmte Teile der Hardware. Gegenstand der Migration ist stets ein gesamtes IS, wobei nicht alle Komponenten tatsächlich einer Veränderung unterworfen werden müssen. Jede Komponente des Quellsystems wird vollständig oder teilweise

1. unverändert in das Neusystem übernommen (vollständiger Investitionsschutz),
2. mit qualitativen Änderungen in das neue IS überführt (partieller Investitionsschutz),
3. durch eine neue - individualerstellte oder als Standardlösung übernommene - Komponente ersetzt (Innovation) oder

4. im neuen System ersatzlos gestrichen, sofern die Funktionalität der Komponente im Neusystem nicht mehr benötigt wird (Elimination).

Der Gesamtmigrationsbereich umfaßt alle Komponenten des Altsystems, die nicht unverändert in das Neusystem übernommen werden, d.h. einer qualitativen Überarbeitung unterliegen, ersetzt oder gestrichen werden. Die Übernahme von Altsystem-Komponenten kann mittels eines qualitativ überarbeitenden Reengineerings (bzw. einer Alteration), einer Konversion [Wol83] oder einer Kapselung (auch: *Wrapping*) [Sne96] erreicht werden. Zusammen mit den Verfahren der Innovation (Neuerstellung oder Übernahme einer Standardlösung) bilden diese Methoden die fünf Migrationstypen. Das Schema der Migration zeigt Abbildung 2.

3 Besonderheiten der Migration

Migrationen sind umfangreiche Vorhaben außerhalb des Tagesgeschäftes der IV, die demzufolge Projektcharakter haben. Die Migration unterscheidet sich von anderen IV-Projekten vor allem hinsichtlich der Komplexität. Verstehen wir ein Projekt als System, so wird dessen Komplexität durch die folgenden Faktoren bestimmt: [Reiß95]

1. Vielzahl: Die Zahl der Elemente des Projektsystems und ihrer Beziehungen
2. Vielfalt: Grad der Verschiedenartigkeit der Elemente und Beziehungen
3. Vieldeutigkeit: Unbestimmtheiten, Unsicherheiten, Risiken, Intransparenzen
4. Veränderlichkeit: Dynamische Entwicklung der Elememente und Beziehungen im Zeitablauf

Da eine Migration notwendigerweise die Übernahme von Elementen des Altsystems impliziert, wird die *Vielzahl* von Elementen des Projektsystems nicht - wie z.B. bei der Neueinführung eines IS - nur durch das Neusystem beeinflußt, sondern zusätzlich auch durch das Altsystem und den Weg des Übergangs vom Alt- auf das Neusystem. Die Verschiedenartigkeit der Elemente ist groß, weil sich Alt- und Neusystem technisch und/oder funktional wesentlich unterscheiden. Die *Vielfalt* führt häufig zu Problemen, weil i.d.R. unterschiedliche Personengruppen über das jeweilige IS-Know-how verfügen. Die *Vieldeutigkeit* ist groß, da Wissen über das Altsystem i.d.R. schwer verfügbar ist (mangelnde Dokumentation, Entwicklungspersonal steht nicht mehr zu Verfügung) und mit dem Neusystem häufig eine technologisch neue Welt erschlossen werden muß. Am schwierigsten wiegt jedoch das Problem des Übergangs, da dieser aufgrund der über lange Zeit gewachsenen spezifischen Charakteristika des Altsystems häufig keine Standardmethoden zuläßt und demzufolge hier - auch für Externe - Neuland betreten wird. Der Grad der *Veränderlichkeit* ist tendenziell hoch, da das Altsystem während der kompletten Projektlaufzeit nicht eingefroren werden kann und aufgrund der großen Innovationsgeschwindigkeit auch marktbezogene Komponenten des Neusystems permanent weiterentwickelt werden.

Über die technische Dimension hinaus wird die Projektkomplexität auch durch organisatorische, wirtschaftliche und personelle Faktoren beeinflußt. Hier zeigt sich bei der Migration zunächst keine grundsätzlich höherere Komplexität als bei anderen Informatik-Projekten. Diese Dimensionen werden jedoch selbst durch die technische Dimension der Migration beeinflußt. Die hohe technische Komplexität impliziert die Notwendigkeit besonderer (komplexer) organisatorischer, wirtschaftlicher und personeller Maßnahmen.

4 Gang der Untersuchung

Die Studie wurde als qualitative Untersuchung durchgeführt. Sie basiert auf dem Konzept der am MIT entwickelten Methodik der kritischen Erfolgsfaktoren (CSF-Methode) [Rock79]. Die Ergebnisse wurden aus Expertengesprächen mit Personen gewonnen, die sich aus beruflichen Gründen mit Fragen des Migrationsmanagements befassen. Die Experten gliedern sich in externe Berater einerseits und IT-Verantwortliche bzw. Projektleiter andererseits. Befragt wurden Experten aus insgesamt sechzehn Unternehmen aus den Bereichen Industrie, Dienstleistung und öffentliche Verwaltung.

Die Erfolgsfaktoren wurden jeweils über mehrere Interviewrunden mit den einzelnen Fachleuten konkretisiert und mit den Befragungsergebnissen anderer Experten abgeglichen. Dabei ergab sich ein weitgehend einheitliches Bild über potentielle Erfolgsfaktoren der Migration. Die Ergebnisse werden im folgenden in Kurzform und aufgrund des qualitativen Charakters der Untersuchung in Form von Leitsätzen dargestellt.

5 Die Erfolgsfaktoren des Migrationsmanagements

5.1 Klassifikation von Erfolgsfaktoren

Erfolgsfaktoren (EF) der Migration sind Einflußgrößen und Bedingungen, die für den Erfolg eines Migrationsprojektes bestimmend sind. Die EF lassen sich bezüglich der folgenden Dimensionen unterscheiden:

I. Beeinflußbarkeit durch das Management

   A. Rahmenbedingungs-EF [geringe Beeinflußbarkeit]: Faktoren, die zwar langfristig vom Management des Unternehmens beeinflußt werden können, aber nicht direkt durch das Projektmanagement

   B. EF i.e.S. [große Beeinflußbarkeit]: Faktoren, die vom Projektmanagement direkt beeinflußt werden können

II. Nach der Relevanz/Erfolgswirksamkeit/Kritikalität (Relevanzkriterium)

   A. Nichtkritische EF [geringe Relevanz]

   B. Kritische EF [große Relevanz]

Die empirische Analyse ergab insgesamt vier Rahmenbedingungs-EF und 26 EF i.e.S. Neben den EF selbst sind auch die Beziehungen zwischen ihnen interessant. Die EF beeinflussen sich gegenseitig hinsichtlich ihrer inhaltlichen Ausprägung und ihrer Relevanz.

Während die Beeinflußbarkeit durch das Management im wesentlichen objektiv festgestellt werden kann, ist die Relevanz von EF offensichtlich hochgradig von der spezifischen Projektsituation abhängig. Ein generelles Urteil über die Relevanz von EF kann aufgrund der Komplexität und der folglichen Vielschichtigkeit der Migrationssituation nicht ermittelt werden. Die EF bekommen erst situativ die Bedeutung kritischer EF.

5.2 Rahmenbedingungs-Erfolgsfaktoren

Als vom Management der Migration nicht direkt beeinflußbare EF wurden festgestellt:

- **EF 1: Existenz von Unternehmensstrategie und Informatikstrategie sowie deren Grad an Harmonisierung:** Die Existenz einer richtungsweisenden Informatikstrategie, die Handlungsrichtlinien für die langfristige Planung der IT umfaßt, fördert das rechtzeitige Erkennen von Migrationsbedarf und erleichtert die Bestimmung der Migrationsstrategie. Projekte, die ohne die Vorgaben einer Informatikstrategie durchgeführt wurden, benötigten signifikant höhere Anlaufzeiten und häufig auch mehrere Anlaufversuche. Der Grad der Harmonisierung zwischen Unternehmensstrategie und Informatikstrategie beeinflußt maßgeblich den Nutzen, den die IV im allgemeinen und die Migration im speziellen für das Geschäft stiften kann. Zwischen Unternehmens- und Informatikstrategie muß ein interdependentes Verhältnis entstehen.
- **EF 2: Spezifische technische und wirtschaftliche Eigenschaften des Altsystems bzw. des Gesamtmigrationsbereiches:** Die spezifischen Eigenschaften bestimmen maßgeblich über Machbarkeit, Dringlichkeit und den Schwierigkeitsgrad der Migration. Zu den spezifischen Eigenschaften gehören u.a. der Umfang des Gesamtmigrationsbereiches, die funktionale und technische Qualität sowie die wirtschaftliche Bedeutung des Altsystems, die Schnittstellen des Altsystems nach außen und dessen Zerlegbarkeit in Komponenten mit eindeutigen Schnittstellen („*Decomposability*") [BrSt95].
- **EF 3: Migrationsaffinität der Unternehmenskultur:** Das Konzept der Unternehmenskultur versucht die eher *irrationalen* Aspekte im Verhalten von

Mitarbeitern zu erklären. Aus diesem Grund ist die Unternehmenskultur nur schwer definier- und meßbar. Nach der der Einschätzung der befragten Experten begünstigen Charakteristika der Unternehmenskultur über eine Ausprägung innovativer Einstellungen die Unterstützung des Projektes durch das Top-Management, erleichterten die Schaffung eines positiven Projektklimas und wirkten fördernd auf den Grad der Projektunterstützung durch die Anwender bzw. die Fachabteilungen. Bei eher konservativen Unternehmenskulturen ist mit tendenziell größeren Widerständen zu rechnen.

- **EF 4: Verfügbare Erfahrungen/Qualifikationen in bezug auf das anstehende Migrationsprojekt:** Eine Migration ist um so leichter durchführbar, je mehr Erfahrungen mit der spezifischen Migrationssituation (qualitativ und quantitativ) existieren. Die gemachten Erfahrungen müssen verfügbar sein (als Dokumentation oder „Wissen" der Mitarbeiter). Die verfügbaren Erfahrungen sind nicht nur von Umfang und Beschaffenheit früherer Migrationsprojekte abhängig, sondern auch von deren systematischer Auswertung.

5.3 Erfolgsfaktoren i.e.S.

5.3.1 Erfolgsfaktoren des Analysemanagements

Dieser Prozeß subsumiert alle Analyse-Aktivitäten, die sich auf das Altsystem und relevante Marktentwicklungen beziehen. Es wurden die folgenden EF ermittelt:

- **EF 5: Analyse der spezifischen Eigenschaften des Altsystems und seiner Komponenten:** Je größer die über entsprechende Analysen hergestellte Transparenz über die spezifischen Eigenschaften des Migrationssystems, desto besser sind die Voraussetzungen für ein erfolgreiches Migrationsprojekt. Bei der funktionalen Analyse sind die Fachbereiche einzubeziehen. Voraussetzung für die Bestimmung der funktionalen Qualität und der strategischen/operativen Bedeutung ist die Kenntnis des Anforderungsprofils der Fachseite.
- **EF 6: Analyse relevanter Marktentwicklungen:** Die Entscheidungen bezüglich der Migrationsstrategie sind wesentlich von der erzielbaren Markttranspa-

renz über verfügbare Systeme/Komponente und Werkzeuge abhängig. Darüber hinaus sind allgemeine Entwicklungsrichtungen der IT zu bestimmen, um Innovationspotentiale auszuschöpfen und auf absehbare Marktentwicklungen vorbereitet zu sein.

5.3.2 Erfolgsfaktoren der Projektinitiierung

Der Prozeß der Projektinitiierung umfaßt alle Aktivitäten, die bis zum eigentlichen Start des Projektes erfolgen.

- **EF 7: Management der Erkenntnis von Migrationsbedarf:** Die Analyseergebnisse bzgl. der spezifischen Eigenschaften des Altsystems und relevanter Marktentwicklungen müssen ggf. frühzeitig in Projektinitiierungsmaßnahmen umgesetzt werden. Anlaß einer Migration kann Problemdruck aufgrund funktionaler oder technischer Faktoren sein (alternativ: "*technology push*" vs. "*demand pull*"). Weiterhin kann zwischen unternehmensinternen und -externen Wandlungsimpulsen unterschieden werden.
- **EF 8: Durchführung von Vorstudien: Machbarkeitsstudie und Kosten-/ Nutzenanalyse als Entscheidungsbasis für das Management:** Die Qualität der gesamten Migrationsplanung basiert auf dem Wissen über die Ausgangssituation und mögliche Zukunftsszenarien. Je früher über Machbarkeitsstudien Planungsfehler erkannt werden, desto besser können deren Folgen begrenzt werden. Elementarer Bestandteil der Initiierungsphase ist die Untersuchung von Kosten und Nutzen. Bei den untersuchten Projekten wurde die Kosten-/ Nutzen-Analyse in zahlreichen Fällen vernachlässigt; fast ausnahmlos kam es dann aber während der Projektdurchführung zu Diskussionen über Projektaufwand und -nutzen, die die Projektarbeit negativ beeinflußten.

5.3.3 Erfolgsfaktoren in Zusammenhang mit der Migrationsstrategie

Die Migrationsstrategie legt einerseits das Ziel der Migration fest und bestimmt andererseits, *wie* dieses Ziel erreicht werden soll.

- **EF 9: Projektzielmanagement:** Das fachlich-inhaltliche Projektziel setzt sich zusammen aus Umfang und Struktur des Gesamtmigrationsbereichs (*was* soll migriert werden?) und strategischer Zielrichtung der Migration (*wohin* soll migriert werden?). Das Projektziel ist zentraler Bestimmungsfaktor der Projektkomplexität; es ist so zu bestimmen, daß Quell- und Zielumgebung beherrscht werden. Das Projektziel muß positiv mit den persönlichen Zielen der Mitarbeiter und Betroffenen korrelieren, ansonsten besteht Konfliktgefahr. Das Projektziel ist hinreichend präzise zu formulieren. Es kann während der Projektdurchführung schrittweise konkretisiert werden, darf aber nur im Ausnahmefall größeren Veränderungen unterworfen werden (Problem der "*Moving Targets*"). Es sind Abhängigkeiten, wechselseitige Beeinflussungen und Konflikte der einzelnen Ziele zu analysieren.
- **EF 10: Strukturierung des Gesamtmigrationsbereichs und Schnittstellenanalyse:** Die Strukturierung des Gesamtmigrationsbereiches in separate Migrationsbereiche ist Voraussetzung für die Definition von Arbeitspaketen, die Ermöglichung von Arbeitsteilung im Projekt, die Implementierung einer verteilten Architektur [GaBr95] und eine Umstellung in mehreren einzelnen Schritten. Die Strukturierbarkeit ist abhängig von der Soft- und Hardwarearchitektur des Altsystems, auf die jedoch durch ein vorbereitendes Reengineering noch Einfluß genommen werden kann. Die Schnittstellen zwischen den Migrationsbereichen sowie zu externen Komponenten sind detailliert festzuhalten. Sie haben maßgeblichen Einfluß auf die Projektkomplexität und den technischen und organisatorischen Koordinierungsaufwand.
- **EF 11: Bestimmung des Vorgehensmodells:** Je größer die Projektkomplexität der Migration, desto mehr raten die Experten grundsätzlich zu einer evolutionären bzw. iterativen Ausgestaltung des Vorgehenskonzeptes. Bei umfangreichen fachlichen Veränderungen und einer entsprechenden Partizipation der Anwender/Fachseite wird ein Prototyping-orientierten Vorgehen empfohlen. Je komplexer sich das Migrationsprojekt darstellt, desto aufwands- und zeitintensiver muß sich die Testphase gestalten. Die Erfahrungen zeigen, daß der Test mehr als die Hälfte der Projektzeit in Anspruch nehmen kann.

- **EF 12: Timing (Zeitpunkt Projektstart, Meilensteine, Migrationsdauer):** Während eine zu frühzeitig eingeleitete Migration überflüssige Aufwendungen verursacht, gefährden verschleppte Migrationen die Einsatzbereitschaft des Migrationssystems und den reibungslosen Verlauf des Migrationsprojektes (zu hohe Dringlichkeit, zunehmende Systemkomplexität). Durch stark verzögerte Projektstarts war bei einigen untersuchten Projekten der Zeitdruck so groß, daß die Qualität des Projektergebnisses in Mitleidenschaft gezogen wurde.

- **EF 13: Festlegung der Migrationstypen:** Der erreichbare Nutzengewinn ist durch die Einführung neuer Komponenten höher als bei einem Reengineering und dies wiederum höher als bei einer Konversion oder Kapselung. Ebenso verhalten sich jedoch auch die Kosten. Die Balance zwischen Innovation und Investitionsschutz ist eine besondere Herausforderung der Migration. Es ist in diesem Zusammenhang intensiv zu testen, welche Altsystem-Komponenten sinnvollerweise in das Neusystem übernommen werden können.

- **EF 14: Einsatz von Migrationsinstrumenten:** Um ein Migrationsprojekt kalkulierbar zu machen und den Kosten- und Zeitaufwand in vertretbaren Grenzen zu halten, sollte die Umstellung weitgehend durch Werkzeuge unterstützt werden. Wichtig ist ein hoher Automatisierungsgrad vor allem dort, wo der erreichbare Nutzen eher gering ist, also bei der Konversion und Kapselung. Die Leistungsfähigkeit verfügbarer Werkzeuge ist heute beschränkt; fast immer waren bei den untersuchten Projekten zusätzliche manuelle Eingriffe erforderlich, da die Werkzeuge mit der Komplexität der spezifischen Migrationssituation überfordert sind.

- **EF 15: Bestimmung der Produktivsetzungsmethodik:** Nachdem die Zielkomponenten erstellt wurden, kommt es zur Übernahme produktiver Aufgaben im Neusystem. Die Übergangsphase ist die kritischste Phase der Migration, insbesondere wenn das zu migrierende System *mission-critical* ist und die stetige Einsatzbereitschaft sichergestellt werden muß. Dabei sind die folgenden Möglichkeiten denkbar:

  1. Sofortiger Umstieg ("*Big Bang*"): Der Übergang auf das Neusystem erfolgt

"mit einem Schlag"; unmittelbar nach der Produktivsetzung des Neusystems wird das Altsystem außer Kraft gesetzt. Dies ist grundsätzlich die kostengünstigste Alternative, aber die mit dem höchsten Risiko.

2. Parallele Migration: Bei der parallelen Migration erfolgt ebenfalls die Produktivsetzung des Neusystems auf einmal, aber Alt- und Neusystem arbeiten über einen bestimmten Zeitraum vollständig *produktiv* parallel, so daß Ausfälle des Neusystems das operative Geschäft nicht gefährden. Für die Systembetreuer oder die Anwender verdoppelt sich häufig der Aufwand, deshalb sollte die Parallelzeit so kurz wie möglich gehalten werden. Dies ist die teuerste Methode, aber auch die sicherste.

3. Schrittweise Migration über einen längeren Zeitraum: Die Funktionalität wird in Paketen übertragen. Diese Methode ist sehr zeitintensiv, aber bei komplexen Migrationen häufig die einzig durchführbare. Die schrittweise Migration ermöglicht es, kurzfristige Ergebnisse ("*Early Wins*") sichtbar zu machen. Dies ist bei langen Projektlaufzeiten wichtig, um den Projekterfolg negativ beeinflussende Verhaltensweisen bei den Beteiligten und Betroffenen (aufgrund von Resignation, Enttäuschung, Motivationseinbruch, etc.) zu vermeiden. Zu beachten ist, daß bei der schrittweisen Migration Altkomponenten zeitweise mit anderen Neukomponenten interagieren und entsprechende Lösungen für die Kommunikation gefunden werden müssen (z.B. *Gateways*) [BrSt95].

5.3.4 Erfolgsfaktoren der Koordination

- **EF 16: Projektexternes Schnittstellenmanagement, Koordination mit Transformationen über das Migrationsprojekt hinaus:** Die Informatik ist kein Selbstzweck und kann Nutzen vor allem als "*Enabler*" für Veränderungen in anderen Bereichen generieren. Entsprechend muß die Migration mit Umstellungen in andere Bereichen verbunden werden. Dies kann sich auf einzelne Geschäftsprozesse oder eine komplette Transformation einschließlich einer strategischen Neuorientierung beziehen [Park96]. Während in der Theorie häufig das Motto "Organisation vor Technik" propagiert wird, läßt sich in der Praxis beobachten, daß nicht selten erst aufgrund einer geplanten Migration Geschäftsprozeßoptimierungen durchgeführt werden; dieses Phänomen ist bei vielen SAP R/2-R/3-Migrationen zu beobachten. Die Studie ergab, daß nicht die Frage des Auslösers von Veränderungen, sondern die der Ausrichtung aller Veränderungen auf die Gesamtstrategie des Unternehmens entscheidend ist.

- **EF 17: Projektinternes Schnittstellenmanagement:** Der Abgleich der Ergebnisse der Teilprojekte ist elementar für den Projekterfolg. Laufender Koordinierungsbedarf entsteht durch Änderungen der Anforderungen durch das Projektzielmanagement. Die Projektschnittstellen nach innen und nach außen sollten weitgehend durch konkrete Vereinbarungen geregelt werden.
- **EF 18: Technische Koordination:** Die Weiterentwicklung der sich noch im Produktiveinsatz befindlichen Altkomponenten muß während der Migration grundsätzlich eingefroren werden. Nur Notwartung darf erlaubt sein. Beim Paralleleinsatz von alter und neuer Welt ist zu beachten, daß Alt- und Neukomponente gleichermaßen bedient werden, damit die in den verschiedenen IS erzielten Ergebnisse identisch sind.

5.3.5 Erfolgsfaktoren des allgemeinen Projektmanagements

Dem allgemeine Projektmanagement sind diejenigen Aktivitäten zugeordnet, die nach Einschätzung der befragten Experten zwar erfolgswirksam sind, aber geringen spezifischen Bezug zum Problem der Migration aufweisen und darüberhinaus nicht das personelle Projektmanagement betreffen. Sie werden hier nur in Kurzform genannt:

- **EF 19: Aufbauorganisatorische Festlegungen: Arbeitsteilung und Kompetenzverteilung**
- **EF 20: Qualitäts-, Termin- und Ressourcenplanung**
- **EF 21: Projektcontrolling**

- **EF 22: Risikomanagement**

5.3.6 Erfolgsfaktoren der personellen Implementierung

Das personelle Projektmanagement umfaßt alle Aktivitäten, die sich primär auf das beteiligte und betroffene Personal der Migration beziehen. Migrationen werden in der Praxis häufig als rein technologisches Problem angesehen. In der Konsequenz können sich erhebliche Widerstände innerhalb des Unternehmens aufbauen, die Migrationen verzögern oder sogar zum Scheitern bringen können.

- **EF 23: Interessengruppenausgleich:** Für eine erfolgreiche Migration müssen die Sichten aller Parteien, die den Projekterfolg mittelbar oder unmittelbar beeinflussen, berücksichtigt werden. Eine Ignorierung von Interessengruppen (z.B. Geschäftspartner, *Shareholder*, Top-Management, Anwender, IT-Spezialisten) erhöht das Erfolgsrisiko. Die volle Unterstützung ist nur gewährleistet, wenn jede Partei einen Nutzen aus dem Projekt ziehen kann. Wenn dies eigentlich unmöglich ist (z.B. Personen werden nachhaltig in ihrer Macht eingeschränkt), muß ein entsprechender Leidensdruck geschaffen werden, so daß die Betroffenen die Lösung immer noch als bestmögliche empfinden. Hier ist ein besonders sensibles Management gefordert.

- **EF 24: Top-Management-Unterstützung:** Es ist notwendig, daß sich eine Führungskraft in verantwortlicher Position zum Sponsor des Projektes erklärt. Es bedarf eines starken, sichtbaren, kontinuierlichen Engagements des Top-Managements im gesamten Projekt. Allerdings darf es nicht unkoordiniert in die Projektarbeit eingreifen.

- **EF 25: Informations- und Kommunikationspolitik:** Migrationsprojekte haben aufgrund ihrer Eigenschaft (komplex, neuartig, riskant) einen hohen Erklärungsbedarf. Eine aktive Informationspolitik (laufende Information der Mitarbeiter über Projektsinn-/-motiv/-ziele/-strategien, Sachfragen, Vorgehensweise und laufende Aktivitäten im Projekt und in bezug zum Projekt) wirkt der Entstehung von Innovationswiderständen entgegen. Die Informationen müssen verständlich, umfassend und offen sein sowie rechtzeitig weitergegeben werden, um Gerüchte zu vermeiden.

- **EF 26: Partizipation der Anwender/Fachseite:** Die Fachseite bzw. der Fachbereich ist der eigentliche Kunde der Migration. Es sollte ihr die Möglichkeit gegeben werden, sich in gestalterischer Rolle verantwortlich im Projekt einzubinden. Die Partizipation von Betroffenen, auch potentieller Opponenten, im Entscheidungsprozeß führt zur Reduktion von Widerständen. Ein hoher Partizipationsgrad trägt zur Motivation und Akzeptanz der Beteiligten und Betroffenen bei. Die Vorteile einer Konsenssuche für die Motivation sind jedoch gegen die Nachteile einer länger dauernden Konfliktlösung abzuwägen.

- **EF 27: Personalselektion, Teamausbildung, Bestimmung struktureller Teammerkmale, Einbindung externer Ressourcen und Vertragsmanagement:** Es sollten für das Team Mitarbeiter mit Migrationserfahrung, mit sozialen Fähigkeiten, intrinsischer Motivation und starkem Selbstvertrauen selektiert werden. Die Kontinuität der Mitarbeiter im Projekt ist sicherzustellen; hier kam es bei den untersuchten Projekten häufig zu Konflikten mit der Fachseite. Für die Teamausbildung ist ein Schulungskonzept erforderlich. Die Arbeit im Team verlangt mindestens 75% der Arbeitszeit einer Person, idealerweise 100%. Da der Neuigkeitsgrad von Migrationen meist hoch ist, kann i.d.R. nicht auf externe Berater verzichtet werden.

- **EF 28: Teamführung und Kooperationsbereitschaft sowie Motivations- und Kontaktfähigkeit des Projektleiters:** Die Teamführung und Kooperation ist die zentrale Aufgabe des Projektleiters. Er nimmt als Motivator, Koordinator und Lenker der Migration die Kernposition der Projektorganisation ein. Die allgemeine Kontaktfähigkeit des Projektleiters hat herausragende Bedeutung für das Image des Projektes und die Durchsetzung der Projektziele. Mit der Person des Projektleiters wird das Projekt selbst assoziiert.

- **EF 29: Ausgestaltung des Anwenderservice (Anwenderschulungen und ad-hoc-Service/Trouble Shooting):** Es wird empfohlen, die Schulungsmaßnahmen in einem Schulungskonzept systematisch zu planen und zu dokumentieren. Insbesondere Inhalt und Zeitpunkt der Schulungen stellen erfolgskritische Faktoren dar. In der Praxis werden oft zu allgemeine oder abstrakte Inhalte geschult und darüber hinaus zu früh oder zu spät mit den Schulungen begonnen. Ein ad-hoc-Benutzerservice muß dafür sorgen, daß nach der Einführung eine regelmäßige Begleitung der Anwender stattfindet.

- **EF 30: Konfliktmanagement:** Konflikte werden häufig als das größte Problem der Teamarbeit bezeichnet. Zur Konfliktlösung ist grundsätzlich ein situativer Führungsstil angezeigt, der aktive und frühzeitige Konfliktbewältigung betreibt und latenter Projektkonflikte nicht verdrängt. Zur Konfliktlösung sind entsprechende Eskalationsprozeduren zu etablieren.

6 Schlußbemerkung

Im Zusammenhang mit dem Einfluß der IT auf die Effektivität und Effizienz aller Unternehmensaktivitäten stellt die Migrationsfähigkeit von IS eine Eigenschaft mit strategischer Bedeutung dar. Permanenter Wandel gehört heute zum unternehmerischen Alltag - dies gilt insbesondere für die Informatik. Erfolgreiches Migrationsmanagement setzt die Beachtung bestimmter Erfolgsfaktoren voraus. Schon im Vorfeld einer Migration ist jedoch grundsätzlich zu bedenken, daß die Innovationen von heute die Migrationsobjekte von morgen darstellen. Deshalb muß auf die Migrationsfähigkeit eines IS bereits bei dessen Neuerstellung geachtet werden. Gesichert wird dies durch die konsequente Nutzung von Standards, Strukturierungs- und Modularisierungsmethoden und offenen Systemkomponenten.

Literatur

[BrSt95] Brodie, M. L.; Stonebraker, M.: Migrating Legacy Systems: Gateways, Interfaces and the incremental Approach, San Francisco, 1995

[Döm97] Dömer, F.: Migration in der Informatik: Die schwierige Balance zwischen Investitionsschutz und Innovation, in: HMD 194, 1997, S. 6-23

[GaBr95] Ganti, Narsim; Brayman, William, The Transition of Legacy Systems to a Distributed Architecture, John Wiley & Sons, New York, 1995

[Park96] Parker, M. M., Strategic Transformation and Information Technology: Paradigms for Performing While Transforming, Prentice-Hall, 1996

[Reiß95] Reiß, Michael: Komplexitätsmanagement, in: Corsten, H. (Hrsg.): Lexikon der Betriebswirtschaftslehre, 3. Aufl., München/Wien, 1995

[Rock79] Rockart, John F.: Chief executives define their own data needs, in: Harvard Business Review, No. 2, 1979, S. 81-93

[Sne96] Sneed, H. M., Die Einbindung alter Host-Software in eine Client/Server-Architektur, in: Objektspektrum, Ausg. 4, 1996, S. 36-43

[Wol83] Wolberg, J. R.: Conversion of Computer Software, New York, 1983

Das Jahr 2000: Vorgehensweisen für die Konvertierung zweiziffriger Jahresdarstellungen in betrieblichen Informationssystemen

Knut Hildebrand

Abstract

Ziel dieses Artikels ist es, das Problem der zweiziffrigen Speicherung von Jahresdaten in seiner Dimension eingehend zu erläutern und Lösungsansätze aufzuzeigen. Von zentralem Interesse ist dabei ein methodisches Vorgehen, welches in dem vorgestellten Modell auf der Basis der Portfolioanalyse erfolgt. Die erforderlichen Schritte werden eingehend diskutiert, unter Berücksichtigung ihrer informationstechnischen Relevanz.

1 Betriebliche Informationssysteme und das Jahr 2000

Vor dem Schritt ins 21. Jahrhundert, welches am 1.1.2001 beginnt, müssen die Informationssysteme eine Prüfung bestehen: den 1. Januar 2000. Konkret: Verkraften die Systeme die neue Jahreszahl, oder gibt es Probleme in den Darstellungen und Algorithmen, weil das Jahr *nicht* mit vier Ziffern abgebildet wurde, sondern nur mit zwei Stellen?

Die Ursache des eigentlich überflüssigen Phänomens ist in den 60er bis 80er Jahren zu finden, als es darum ging, Erfassungsaufwand und teuren Platz zu sparen im Haupt- und Plattenspeicher, indem auf zwei Ziffern in der Jahresdarstellung – Jahrhundert und Jahrtausend – verzichtet wurde. Aber selbst heute findet sich in aktuellen Konfigurationen das gleiche Problem, obwohl die Restriktionen weitgehend aufgehoben sind.

Die Evolution betrieblicher Informationssysteme bzw. Datenbanken/Datenfelder ist keine überraschende Tatsache. Vielmehr finden sich praktisch permanent Beispiele für die Anpassung an neue Umweltbedingungen oder betriebliche Erfordernisse. In der Vergangenheit waren dies etwa die geplante Einführung der „Quellensteuer“ oder die Umstellung der Postleitzahl am 1.7.1993 auf fünf Stellen. Zu erwarten ist in naher Zukunft die europäische Währungsunion, die u.a. ein neues Währungsfeld (EURO) mit sich bringt.

2 Die Dimension des Problems

Das Problem der zweiziffrigen Speicherung der Jahresangabe kann unter mehreren Aspekten betrachtet werden; dazu zählen: die unterschiedlichen *Datumsformate*, die *Orte*, an denen sie zu finden sind, die *Funktionen*, bei denen sie auftreten, aber auch die *Kosten*, die mit einer Korrektur verbunden sind. Veröffentlichungen zu diesem Problem finden sich u.a. auch im Internet ([cnn], [data], [ibm], [Knol96], [Knol97a], [Knol97b], [s390], [Walt97]).

2.1 Datumsformate

Bei den betroffenen Datumsformaten läßen sich eine Vielzahl von Variationen feststellen, von denen hier einige gezeigt werden (Abb. 1). Darüber hinaus können die aufgezeigten Formate ergänzt werden um Trennzeichen: Leerzeichen, Punkt (11.10.96), Bindestrich (96-04-21), Schrägstrich (10/23/96) usw.

TTMMJJ	europäischer Standard
JJMMTT	internationales wissenschaftliches Format
TTTJJ	ohne Monatsangabe (Julian)
MMTTJJ	US-amerikanisches Format
TTMM19JJ	mit festem Jahrtausend/-hundert
TTMM199J	mit festem Jahrtausend/-hundert/-zehnt

Abb. 1: Typische Datumsformate

Neben diesen gebräuchlichen Standardformaten treten auch Fälle auf mit Textkomponente – zum Beispiel: Montag, 16. Juni 1952, 19:52 Uhr (mit oder ohne Zeitkomponente), bzw. mit Datum/Zeitstempel als Textfeld (z.B. im Serienbrief): München, den 30. Okt. 97 – oder mit dem Datum als Teil einer Identifikationsnummern (z.B. als Teil der Rechnungsnummer).

2.2 Vorkommen von Datumsfeldern

Nach der Vielfalt von Datumsformaten mit Jahresangaben ist als nächstes zu klären, wo überall mit ihrem Auftreten zu rechnen ist. Denn nur wenn die Orte bekannt sind, können die Jahresangaben auch modifiziert werden. Charakteristische Stellen in Informationssystemen sind: (Anwendungs-) Programme, Betriebssystem-Prozeduren (BAT, JCL u.a.), Systemzeiten, Systemprogramme, Hilfs- und Dienstprogramme (Sortierer), Werkzeuge (CASE-Tools), Ein-/Ausgabe-Masken, Listen/Reports, Datumsroutinen (Prüfung und Umwandlung), Tabellen (inkl. Index, Subskripte), Dateien (auch temporär), Datenbanken, Programmiersprachen (z.B. mit Deklarationen PIC XX, PIC 99, DATE) und Microcode.

2.3 Funktionen mit Datumsfeldern

Selbst wenn alle Datumsformate und Orte bekannt sind, ist eine wie auch immer geartete automatische Konvertierung vor ein weiteres Hindernis gestellt. Wie, d.h. bei welchen Funktionen – zeitpunkt- und zeitraumbezogenen Fragestellungen [HiMü91] –, werden Datumsfelder berechnet oder verändert, welche Ergebnisse werden damit erzielt? Typische Funktionen sind: Eingaben, Ausgaben, Vergleichen, Sortieren und die Berechnung von Datumsfeldern. Hier können bei der zweiziffrigen Jahresdarstellung Fehler geschehen, wenn beispielsweise ein 1957 Geborener im Jahr 2000 plötzlich 57 Jahre alt ist (00 – 57 = -57, ohne Vorzeichen: 57).

2.4 Kostenaspekte

Hinsichtlich der Kosten lassen sich nach Schätzungen folgende Werte ermitteln ([itaa], [year]): in der mikroökonomischen Dimension, also innerhalb eines Unternehmens, betragen die Kosten ca. 0,2 $ bis 1 $ pro Programmzeile – mit den bekannten Problemen ihrer Messung [Jone87] –, d.h. in großen Unternehmen ca. 50-100 Mio. $. Dies führt in der makroökonomischen Dimension weltweit zu vielen Mrd. $ Aufwand.

3 Das Vorgehensmodell

Die *Durchführung* der Konvertierung der temporalen Informationen sollte in Form eines eigenständigen Projektes erfolgen (Abb. 2).

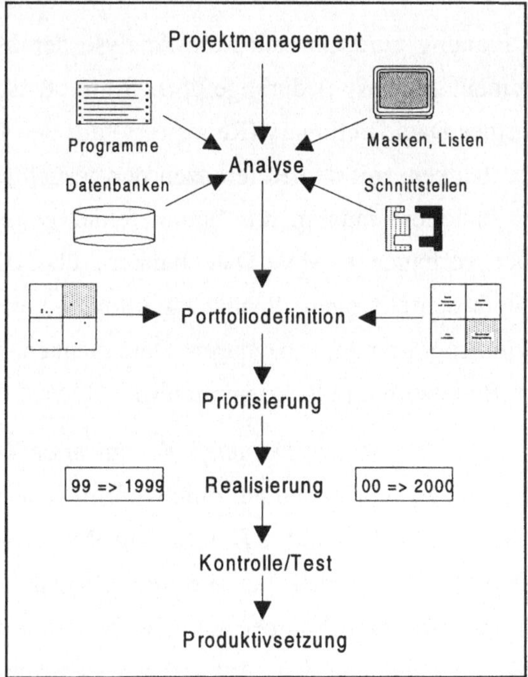

Abb. 2: Das Vorgehensmodell im Überblick

Die *Planung* der Umstellung betrifft im wesentlichen folgende Punkte:

- Ziele
- Voraussetzungen und Restriktionen

- Projektmanagement: Projektstandards, -team, -umfang und -verantwortliche
- Budget und Aufwand
- Ansprechpartner (Hardware, Software, Berater usw.)
- Qualitätssicherung
- Chancen und Risiken
- Methoden und Werkzeuge
- Zeitplanung
- Auswirkungen

3.1 Analyse und Identifikation: Suchen und Finden

Nach Abschluß der Planung wird als erstes die Analyse der Informationssysteme (Software und Dokumentation usw.) durchgeführt. Sinnvoll ist, sofern nicht vorhanden, der Aufbau eines Data Dictionary (Repository) aus den Daten der produktiven Programme/Datenbanken, mit den Referenzen der gesuchten Datenfelder. Um dieses „Inventar" zu erstellen, müssen die Datumsfelder gefunden werden. Dies kann geschehen über vorhandene Meta-Datenbanken, über Copystrecken, Programmcode und Dateibeschreibungen, Search-Funktionen von Editoren, Pattern-Suche mit Wildcard in Programmen, Prozeduren, Datenbanken, Dateien usw. sowie durch Befragung von Benutzern und Programmierern.

Dieser Review zur Identifikation der *Datumsfelder* an allen *Orten* sowie in den *Verarbeitungsregeln* ist dann relativ einfach durchzuführen, wenn alle datumsspezifischen Informationen vorhanden sind, z.B. in einem Repository. Anderfalls muß ein Suchverfahren durchlaufen werden, bis alle Orte abgesucht worden sind. Begonnen werden kann etwa mit dem Sourcecode eines Programms, aus dem alle relevanten Datumsfelder ermittelt werden. Anhand der Referenzen auf programmexterne Felder (logische Dateien, Linkage Section zu anderen Programmen) werden im nächsten Schritt die abhängigen Daten ermittelt, die dann weiter verfolgt werden. Dies können z.B. Reports sein, die diese Datei benutzen, oder andere Programme, die auf die gleiche physische Datei zugreifen. Auf diese Weise kann man sich durch das System arbeiten und alle Informationen ermitteln (Abb. 3).

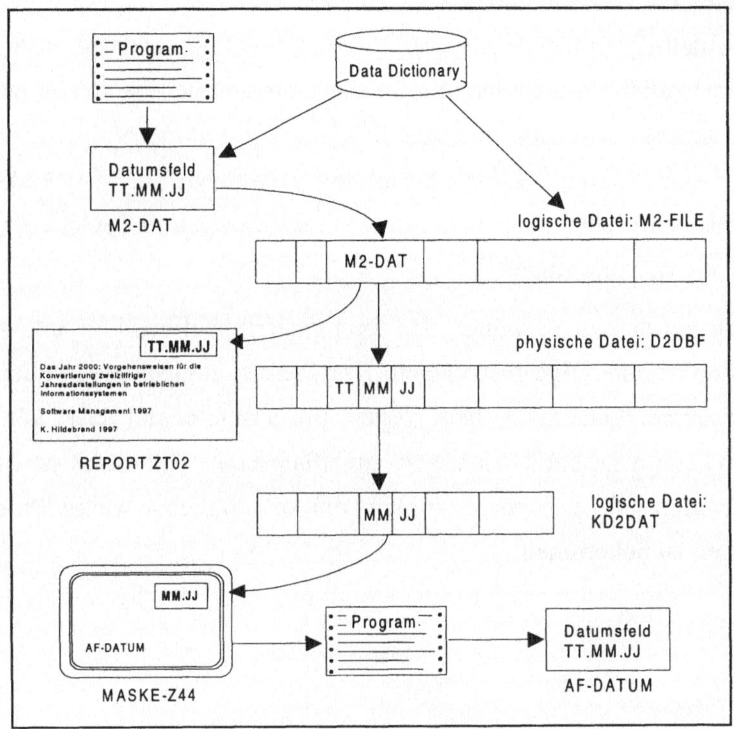

Abb. 3: Identifikation von Datumsfeldern

Auf der Grundlage der ermittelten Daten sind folgende Metriken ableitbar:

- die Anzahl der Datumsfelder (Attribute), davon beschreibend/identifizierend (Schlüssel)
- die Anzahl der Datumsfelder pro Programm, Datei, Datenbank, pro 100 Attribute usw.
- die Anzahl der Berechnungen, Arten der Datumsprüfung, -konvertierungen
- die Anzahl der Beziehungen: Datumsfelder als Schlüssel oder Fremdschlüssel, als Attribut und abgeleiteter Attributwert (Bsp.: Monat und Monat + 3)

3.2 Festlegen der Umstellungsstrategie: Portfoliogestützte Priorisierung

Aus den Daten der Analyse und den Metriken läßt sich jetzt ein Portfolio aufstellen (Abb. 4). In dem Portfolio werden die Programme/Datenbestände zum einen eingeordnet hinsichtlich ihrer Bedeutung für das Unternehmen im Bezug auf das Jahr 2000. Zum anderen wird differenziert, ob keine Umstellung nötig ist, ob die Programme nur eine externe 2-Zifferndarstellung verwenden (z.B. in Masken, Listen oder Schnittstellen), oder ob ebenso intern (in der Logik und Speicherung) vielfältige Jahresformate vorkommen.

Das Portfolio wird dazu verwandt, den Reengineeringaufwand zu analysieren und die Umstellungsreihenfolge festzulegen. Der Priorisierungsprozeß sollte die kritischen Programme (Quadrat 3) bevorzugen, um anschließend das zweite, sechste und fünfte Quadrat zu berücksichtigen. Innerhalb eines Quadrates ist es sinnvoll, schnittstellenminimale Subsets zu bilden, um so möglichst wenig Probleme mit Seiteneffekten zu bekommen.

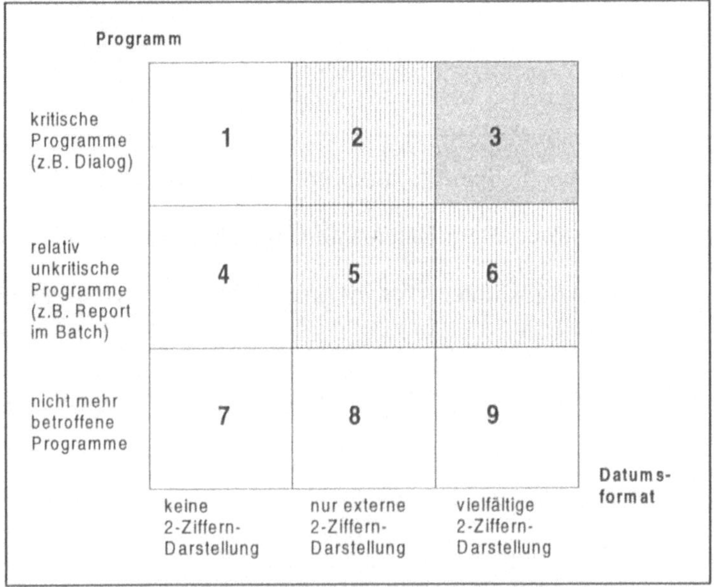

Abb. 4: Datumsformat-Programm-Portfolio

Bei der Definition der Umstellungsstrategie sind einige Spezialfälle hinsichtlich der

Feldmigration/-evolution von zeitbezogenen Daten zu beachten, die unter Umständen die Prioritätensetzung beeinflussen:

- Was geschieht mit archivierten Daten (und deren Auswertungsprogrammen), sollen auch sie migriert werden?
- Temporäre Dateien (Jahresabschluß usw.), die vielleicht nicht zum Umstellungszeitpunkt bestehen, aber trotzdem betroffen sind.
- Gleiches gilt für Archivierungs-, Sortier- und Mischprogramme, die nur gelegentlich benötigt werden.
- Datumsfelder werden auch an andere Programme weitergegeben. Hier existieren unternehmensinterne und -externe (DFÜ, EDI) Schnittstellen, die gegebenenfalls – gerade bei externen Partnern – auch modifiziert werden müssen.
- Feldänderungen bei zeitbezogenen Schlüsselattributen führen zu Schlüsseländerungen, die wiederum, etwa bei einer indexsequentiellen Speicherung (ISAM, VSAM-KSDS), die Reihenfolge der Datensätze beim sequentiellen Zugriff beeinflußt. Damit können Algorithmen, z.B. die Gruppenwechselverarbeitung, fehlerhafte Ergebnisse produzieren; ebenso sind Start, Read Next oder Alternate Key Statements betroffen.
- Das Jahr 2000 ist ein Schaltjahr; nicht alle Datumsprüfungsroutinen beachten dies!
- Aufgrund unsauberer Programmierung kann es vorkommen, daß bestimmte Werte im Jahresfeld eine andere Bedeutung als die Jahreszahl haben. Beispielsweise bedeutet dann der Wert „00“, daß das Feld initialisiert ist, und „99“ ist als Sonderwert (z.B. in Verbindung mit Fehlerroutinen) zu interpretieren.
- Änderungen betreffen nicht nur Variablenvereinbarungen (evtl. mit Value-Klauseln) und prozedurale Teile, sondern auch schwer zu identifizierende Literale (Texte).

3.3 Realisierung: Anpassung der Datenfelder und Verarbeitungsregeln

Im nächsten Schritt ist zu klären, wie alle Datumsfelder an den entsprechenden Orten (Dateien, Programme, Masken usw.) sowie die Algorithmen anzupassen sind an die neue vierziffrige Struktur. Als einheitlicher Standard empfiehlt sich eine Umformatierung und Speicherung des Datums mit acht Stellen, z.B. im Format JJJJMMTT. Der Vorteil liegt darin, daß mit dieser Struktur sehr gut Vergleiche und Sortierungen möglich sind, und daß daraus beliebige Darstellungen – TT.MM.JJJJ, MM.TT.JJ u.a. – abgeleitet werden können.

Wurde vor der Konvertierung das Jahr in zwei Bytes gespeichert, und werden jetzt vier Bytes benutzt, so ergibt sich ein Speicherbedarf von zusätzlichen zwei Bytes pro Feld. Der gesamte Bedarf ergibt sich dann aus der Anzahl der Felder pro Datensatz mal 2, multipliziert mit den der Anzahl der Datensätze (Tupel) pro Datei (Tabelle) über alle Datenbestände.

Bei der prozeduralen Konvertierung sind die Verarbeitungsregeln/Vergleiche anzupassen an die neuen Formate; so werden Fehler bei der Berechnung vermieden, denn in der zweiziffrigen Darstellung ist das Jahr 2000 kleiner als 1999 (00 < 99), was jedoch nicht der Realität entspricht: 2000 > 1999!

Neben dieser aufwendigen, aber klaren und "sauberen" Lösung können noch andere Verfahren gewählt werden, um den Schritt ins Jahr 2000 vorzunehmen. Diese Verfahren sind mit Vorsicht zu genießen, da sie einige Einschränkungen mit sich bringen, haben aber den Vorteil, daß sie intern nur zwei Bytes für die Speicherung des Jahres benötigen. Es handelt sich dabei um zwei Zeitfensterverfahren – *Fixed Window Technique* und *Sliding Window Technique* – und mehrere Methoden zur Datenkompression ([IBM96]):

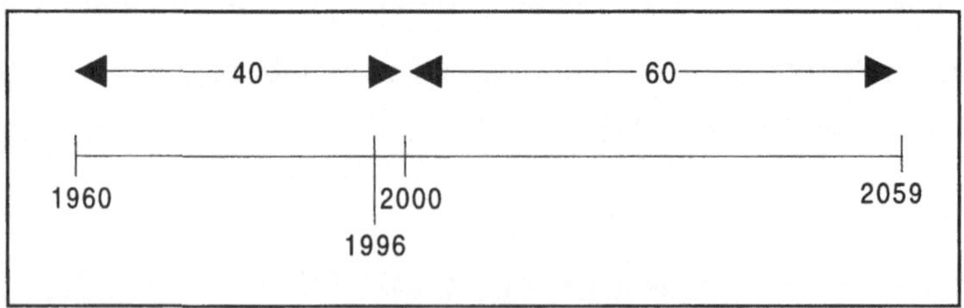

Abb. 5: Fixed Window Technique

Bei der *Fixed Window Technique* wird ein statisches 100-Jahre-Intervall benutzt, das über den Jahrtausendwechsel reicht (Abb. 5). Aus Sicht der Anwendung wird festgelegt, welcher Zeitraum im 20. Jahrhundert liegt, und welcher im 21. Wird beispielsweise als Basisdatum der 1. Januar 1960 gewählt, dann gilt: Werte, die größer oder gleich 60 sind, stellen Jahresangaben dieses Jahrhunderts dar (19xx), andernfalls des nächsten (20xx).

Flexibler ist die *Sliding Window Technique*: der Anwender definiert, welcher Zeitraum immer in der Verganganheit liegen soll, und welcher in der Zukunft (Abb. 6). Durch Manipulation der Systemzeit, die dann von der aktuellen Zeit abweicht, kann ein immer gleichlanges Intervall automatisch fortgeschrieben werden.

Beiden Zeitfensterverfahren ist gemeinsam, daß sie eher für Zwischenlösungen geeignet sind. Sie bringen zwar die Vorteile mit, daß das zweiziffrige Format beibehalten werden kann, und daß daraus trotzdem das richtige Jahrhundert ableitbar ist; die Nachteile aber wiegen schwer: z.B. Probleme, wenn das 100-Jahre-Intervall zu klein wird, Performance-Einbußen, alle Programme müssen die gleichen Algorithmen verwenden, und es gibt massive Probleme, wenn zweistellige Jahreszahlen in Schlüsselfeldern auftreten, z.B. bei Index-Dateien.

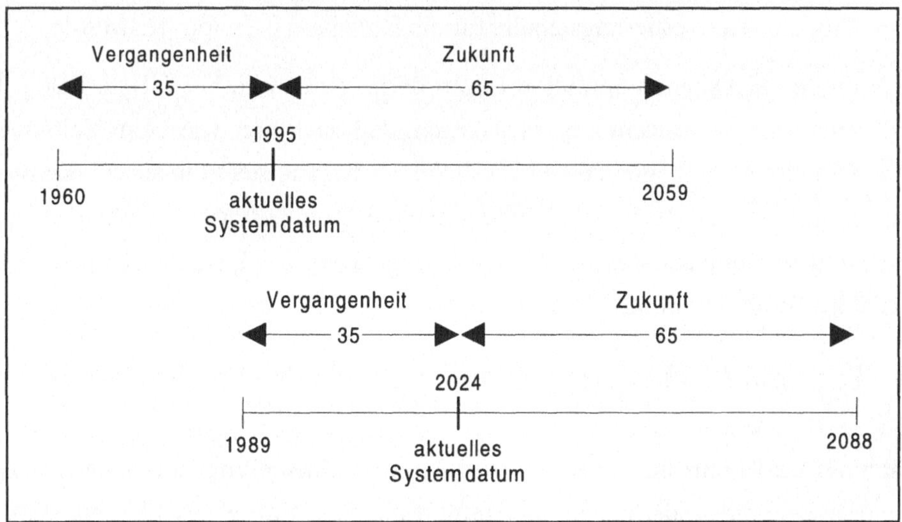

Abb. 6: Sliding Window Technique ([IBM96])

Bei der *Datenkompression* lassen sich mehrere Verfahren unterscheiden, denen allen gemeinsam ist, daß zur Kompression zwei Bytes benötigt werden. Beispielsweise können in einem Byte 2 Dezimalzahlen ohne Vorzeichen gepackt untergebracht werden: 1. Byte X '19', 2. Byte X '78'. Wenn möglich, bietet sich die binäre Darstellung an: in zwei Bytes sind 2^{16} (= 65536) Kombinationen darstellbar.

Oder es werden außer den Ziffern auch die Buchstaben genommen, so daß 36 Zeichen zum codieren zur Verfügung stehen. Wählt man z.B. als Basisjahr den Wert 1900, so berechnet sich das Datum folgendermaßen (Abb. 7): Jahr = $1900_{10} + xx_{36}$, z.B. entspricht die Kombination '1W' dem Jahr 1968 (1900 + $1W_{36}$ = 1900 + 68).

Zweiziffriges Jahr (codiert)	**Konvertiertes Jahr (dezimal)**	**Jahr (mit Basisjahr 1900)**
00 - 0Z	00 - 35	1900 - 1935
10 - 1Z	36 - 71	1936 - 1971
...	...	...
R0 - RZ	972 - 1007	2872 - 2907
...	...	...
Z0 - ZZ	1260 - 1295	3160 - 3195

Abb. 7: Konvertierungstabelle für die Datenkompression ([IBM96])

Die Vorteile der Datenkompression liegen darin, daß weiterhin 2-Byte-Felder benutzt werden können; nachteilig ist dagegen, daß auch hier wie beim Zeitfensterverfahren nur ein eng begrenztes Intervall zur Verfügung steht (außer bei der Binärspeicherung), daß möglichst alle Programme und Datenbestände simultan angepaßt werden müssen, und daß durch die (De-) Kompressionslogik eine zusätzliche Komplexität ins System kommt.

3.4 Test und Produktivsystemumstellung (Produktivsetzung)

Unabdingbare Voraussetzung für eine erfolgreiche Umstellung ist ein eigenes Testsystem, in dem die Änderungen durchgeführt und überprüft werden können. Grundsätzlich finden die bekannten Testverfahren – Programmtest, Integrationstest, Systemtest, Regressionstest usw. – ihre Anwendung ([Myer79], [Somm92]). Darüber hinaus sind spezielle Testverfahren notwendig, die die Datumsproblematik berücksichtigen ([IBM96]):

Hierfür ist als erstes zu beachten, daß bei einer Umstellung der Systemzeit zum Zwecke der Simulation neuer Jahreszahlen unter Umständen bestimmte Funktionen und Ressourcen nicht mehr verfügbar sind, weil ihre Gültigkeitszeit abgelaufen ist. Dazu können zählen: User IDs, Passwörter, Dateien und Datenbanken, Lizenzen, Speichermanagement u.a.m. Beispielsweise könnten alle Dateien, die älter sind als zwei Jahre, gelöscht werden durch automatisierte Prozeduren.

Des weiteren sind Test-Szenarios aufzubauen, die typische Situationen aus dem normalen Betrieb in der "Zukunft" nachbilden. Dazu zählen: tägliche, wöchentliche, monatliche usw. Geschäftsvorfälle/Aktivitäten, der 29. Februar 2000, denn das Jahr 2000 ist ein Schaltjahr, oder der 31.12.2000, um julianische Datumsdarstellungen zu prüfen. Ferner sollten im Testsystem – möglichst ein gespiegeltes Produktivsystem – die verschiedenen Kombinationen von *alter* und *neuer* Zeit, *alten* und *neuen* Daten geprüft werden (Abb. 8).

Systemzeit	**Test-Daten**
aktuell (1997)	vor dem 01.01.2000
aktuell (1997)	nach dem 01.01.2000
31.12.1999	vor dem 01.01.2000
31.12.1999	nach dem 01.01.2000
01.01.2000	vor dem 01.01.2000
01.01.2000	nach dem 01.01.2000
29.02.2000	vor dem 01.01.2000
29.02.2000	nach dem 01.01.2000
31.12.2000	vor dem 01.01.2000
31.12.2000	nach dem 01.01.2000

Abb. 8: Test-Szenarios für konvertierte temporale Daten

Bei der Durchführung der Migration wird das Informationssystem in der vorgegebenen Reihenfolge umgestellt auf die neuen Formate/Prozeduren. Das bedeutet, daß manuelle und automatische Umsetzungsverfahren stattfinden, daß der Abschlußtest durchgeführt wird, und daß alle Aktivitäten entsprechend zu dokumentieren sind. Ferner sind die Anwender- und die Systemdokumentation zu überarbeiten. Nach erfolgreich bestandenem Abschlußtest geht das neue System in Produktion ([IBM96]).

3.5 Abschluß

Zum Abschluß des Projektes entstehen der Abschlußbericht und die Nachkalkulation im Rahmen des IV-Controlling. Aus systemtechnischer Sicht ist Reorganisation der Datenbanken und Dateien sinnvoll, um Performance-Problemen aus dem Wege zu gehen. Ferner ist ein Backup der neuen Strukturen dringend geboten.

4 Resümee

Die Umstellung des Datums von 2 Ziffern in ein Format, das auch im Jahr 2000 eine eindeutige Identifikation des Jahrhunderts zuläßt, ist ein aufwendiges Verfahren, das schon frühzeitig mit den erforderlichen Budgets in Angriff genommen werden muß. Die nötigen Methoden und Werkzeuge existieren bereits und müssen nur noch unternehmensspezifisch innerhalb eines gezielten Vorgehens (Abb. 2) eingesetzt werden.

Dieses Vorgehensmodell ist grundsätzlich auch geeignet für andere Datenfelder und Regeln, wie sie z.B. bei der Einführung der neuen europäischen Währung, des EURO, oder anderer Feldveränderungen denkbar sind. Der Vorteil dieses Modells liegt darin, daß alle Schritte in der erforderlichen Reihenfolge definiert sind, mit ihrem Methoden- und Werkzeugeinsatz; man muß jedoch frühzeitig beginnen, schon allein deshalb, weil immer öfter Planzahlen auftauchen, die das Jahr 1999 überschreiten.

Literatur

[HiMü91] Hildebrand, Knut; Müßig, Michael: Modellierung zeitbezogener Daten im unternehmensweiten Datenmodell (UDM), in: Wirtschaftsinformatik, 33 (1991) 3, S. 238-243

[IBM96] IBM (Hrsg.): The Year 2000 and 2-Digit Dates: A Guide for Planning and Implementation, IBM Form GC28-1251-01, 2. Aufl. 1996

[Jone87] Jones, Capers: Effektive Programmentwicklung – Grundlagen der Produktivitätsanalyse, McGraw-Hill, Hamburg u.a. 1987

[Knol96] Knolmayer, Gerhard: Das Jahr 2000: Eine Herausforderung für das IS-Management, in: Rundbrief Informationssystem-Architekturen des GI-Fachausschusses 5.2, 3. Jg., 1/1996, S. 6-8

[Knol97a] Knolmayer, Gerhard: Das Jahr 2000-Problem: Medien-Spektakel oder Gefährdung der Funktionsfähigkeit des Wirtschaftssystems?, in: Wirtschaftsinformatik 39 (1997) 1, S. 7-18

[Knol97b] Knolmayer, Gerhard: Das Jahr 2000-Problem im Internet, in: Wirtschaftsinformatik 39 (1997) 1, S. 73-76

[Myer79] Myers, Glenford J.: The Art of Software Testing, John Wiley & Sons, New York 1979

[Somm92] Sommerville, Ian: Software Engineering, 4. Ed., Addison-Wesley, Wokingham 1992

[Walt97] Walter, Manfred: Vorgehensweise und Erfahrungen bei der Datumsumstellung im SAP System R/2, in: Wirtschaftsinformatik 39 (1997) 1, S. 19-24

Elektronische Veröffentlichungen (Internet)

[cnn] http://www.cnn.com/TECH/9601/2000/index.html

[data] http://www.datamation.com/PlugIn/issues/1996/jan1/FEATURES.html

[ibm] http://www.software.ibm.com/year2000/paper.html

[itaa] http://www.itaa.org/yr2000bg.htm

[s390] http://www.s390.ibm.com/stories/tran2000.html

[year] http://www.year2000.com/

Konfiguration, Adaption und Wiederverwendung einer Softwarelösung für die öffentliche Verwaltung

Werner Wirdemann

Abstract

Bei dem vorliegenden Beitrag handelt es sich um einen Erfahrungsbericht aus der Praxis des betrieblichen Software Managements, in dem Anforderungen, Lösungskonzepte, Technologien und Methoden zur Konfiguration, Adaption und Wiederverwendung einer Softwarelösung als aktuelles Ergebnis eines laufenden großen Projektes für die Berliner Verwaltung dargestellt werden. Grundidee des Konzeptes ist ein wiederverwendbarer Lösungskern mit verschiedenen Schalen zur Konfiguration und Adaption an die speziell geforderte Variante der Softwarelösung. Im Vordergrund des Beitrages steht dabei die Machbarkeit des Konzeptes und nicht die hohe Originalität der verwendeten Idee. In dem Projekt werden als technologische Grundlage die *Network Computing Architecture* und als methodisches Rahmenwerk der Software-Entwicklungsstandard *V-Modell* eingesetzt.

Einleitung

Ausgehend von einem Pflichtenheft mit speziellen Berliner Anforderungen an eine Verwaltungssoftware hat das Firmenkonsortium Oracle/PSI begonnen, eine Standardlösung für integrierte Gewährung von Verwaltungsleistungen zu entwickeln. Der bundesweite Einsatz und die konsequente Umsetzung einer Standardlösung, die sich flexibel an die Belange der Verwaltung anpassen läßt und sich nahtlos in bestehende Organisationsformen und DV-Landschaften einpassen soll, erfordert eine offene Architektur und Konzepte, die eine flexible Konfiguration der Standardlösung und eine hohe Anpaßbarkeit des Systems an lokale Besonderheiten der einzelnen Städte und ggf. Bezirke ermöglicht. Zusätzlich unterliegen die vom System betroffenen Organisationseinheiten - zumindest in Berlin - brisanten Veränderungen in der Aufbau- und Ablauforganisation und die unterstützten Rechtsgebiete starken Änderungen an Vorschriften und/oder Gesetzen.

Die derzeitig knapper werdenden Finanzmittel der Städte und Kommunen setzen enge Rahmenbedingungen, so daß Konzepte gefordert sind, die auch einen hohen Grad an Wiederverwendbarkeit der für einzelne Dienststellen notwendigerweise individuell erstellten Lösungen vorsehen.

Von einem kleinen Projektteam wurden die Konzepte vorbereitet und eine Untersuchung der Realisierbarkeit unterschiedlicher Softwarearchitekturen durchgeführt. Ein erweitertes Team hat mit der Realisierung des Systems und eines technologischen Piloten begonnen.

Projektdarstellung

Gegenstand des Projektes ist die fachliche Weiter- und technologische Neuentwicklung einer Verwaltungssoftware. Für die Anwender bedeutet die fachliche Weiterentwicklung, daß bisher nicht durch Datenverarbeitung unterstützte Vorgänge automatisiert und interaktiv bearbeitet werden können. Technologische Neuentwicklung bedeutet, eine moderne, leistungsfähige Informationstechnik und Datenverarbeitung effektiv einzusetzen. Der Nutzen des Projektes liegt u.a. in der Verkürzung der Wartezeiten für die Antragsteller und in der schnelleren Erstellung von Bescheiden, um damit einerseits mehr Zeit für die umfassende Beratung der Bürger zu gewinnen und andererseits bessere Arbeitsbedingungen für die Mitarbeiterinnen und Mitarbeiter zu schaffen. Ziel der Software ist es, die Effizienz der Verwaltungsvorgänge zu steigern.

Das Verfahren soll in Berlin an ca. 3000 Arbeitsplätzen in über 70 Dienststellen in derzeit 23 Bezirken eingesetzt werden und ist daher sowohl für die Bürger, die Fachanwender und das Führungspersonal als auch für die Informationstechnologie in der Berliner Verwaltung von großer Bedeutung.

Nach der Erarbeitung eines Projekthandbuches gemäß V-Modell und eines Lastenheftes war die Erstellung eines Pflichtenheftes für das Gesamtprojekt ein wesentlicher weiterer Schritt. An der Erstellung der Anforderungen waren über 50 Mitarbeiter der Senatsverwaltung und der Bezirke beteiligt. Ausgehend von dem 'Berliner Pflichtenheft' hat das Firmenkonsortium Oracle/PSI begonnen, eine neue integrierte Standardlösung zur Unterstützung der Verwaltung zu entwickeln. Ziel des Projektes ist es, für Berlin eine Lösung zu erstellen, die in anderen Bundesländern wiederverwendet werden kann. Das Projekt ist als großes administratives Vorhaben mit über 3000 Function Points und hoher Projektkomplexität eingestuft. Von der Anforderungsanalyse bis zur Einführung werden über 30 Personenjahre veranschlagt und ca. 10 - 15 Personen im Entwicklungsteam eingesetzt.

Anforderungen

Da der Gesetzgeber in einem Bundesgesetz die Grundregeln und -abläufe zur Gewährung von Leistungen der Verwaltung für die entsprechenden Rechtsgebiete weitgehend festgelegt hat, werden Funktionen spezifiziert, von denen von vornherein feststeht, daß andere Städte und Kommunen dieselben oder ähnliche Funktionen benutzen können. Die Gesetze werden in verschiedenen Stellen (z.B. Bundesländern) unterschiedlich durch Ausführungsvorschriften detailliert bzw. interpretiert. Um sich die Flexibilität und die Vielzahl der möglichen Varianten und damit den größtmöglichen Nutzen und die Einsatzmöglichkeit zu erhalten, soll die Lösung - wie bei marktgängigen Standardanwendungen auch - durch Anpassung von Parametern individuell auf die Stadt, den Bezirk bzw. die Kommune zugeschnitten werden. Neben den Anforderungen an eine Integration mit anderen Verwaltungen, objektorientierten Ansätzen und weitgehender Einbeziehung von Standardprodukten ist damit eine größtmögliche Unabhängigkeit und Anpaßbarkeit von der Organisation gefordert.

Zur Lösung dieser grundsätzlichen Anforderungen werden drei zentrale Begriffe betrachtet:

- **Konfiguration** des Systems an jeweilige Besonderheiten der Bezirke und Dienststellen, z.B. Logos im Schriftverkehr
- **Adaption** des Systems an Änderungen in Gesetzen und Organisationen, wobei eine Lösung zwischen folgenden Extremen realistisch ist:

  - der heute leider nicht ganz realistischen Vision: Scannen der Gesetzestexte und die Software „paßt sich automatisch an“ und
  - der heute leider nicht ganz unrealistischen Situation: Veröffentlichung rückwirkender Änderungen der Gesetze mit wochenlangen Programmieraufwendungen.
- **Wiederverwendung** des Systems, um daraus weitestgehend Softwarebausteine für eine bundesweite Standardlösung abzuleiten.

Lösungskonzept

Das im Projekt verwendete Lösungskonzept zur Konfiguration, Adaption und Wiederverwendung der Softwarebausteine ist keine neue oder geniale Erfindung. Die Herausforderung liegt nicht in der Originalität, sondern darin, möglichst einfache Grundideen zu formulieren, allen Projektbeteiligten zu vermitteln und die Grundprinzipien vom Anfang des Projektes bis zur Wiederverwendung der Projektergebnisse methodisch konsequent zu verfolgen sowie eine geeignete Entwicklungs- und Einsatzumgebung zu verwenden. Dieses zu erreichen, ist genau die Aufgabe des *Software Managements*. Eine geeignete Technologie, wie zum Beispiel die Objektorientierung, erleichtert diese Aufgabe, ist jedoch kein Garant für den Projekterfolg. Grundvoraussetzung ist, daß bereits in der Definition und Analyse der fachlichen und organisatorischen Anforderungen die oben genannten zentralen Begriffe berücksichtigt werden. Aus der fachlichen Struktur und Klassifizierung des Anwendungssystems können dann Objekte für einen Lösungskern und zusätzliche Varianten abgeleitet werden.

Lösungskern

Der hier definierte Lösungskern ist die Gesamtheit aller für alle Varianten des Anwendungssystems notwendigen Objekte zur Unterstützung der technologischen Basisfunktionen sowie fachlichen Grundaufgaben, -abläufe und -strukturen.

In dem Lösungskern sind beispielsweise einfache fachliche Grundabläufe einer Vorgangsverarbeitung mit Grundelementen der Aktenverarbeitung, in denen zyklische Abläufe und Bearbeitungszustände frei konfigurierbar sind, enthalten. Typische und immer wiederkehrende Aktivitäten dieser Vorgangsverarbeitung sind zum

Beispiel - ohne Anspruch der Vollständigkeit - bei der Antragsbearbeitung: Erfassen, Ändern, Festsetzen, Prüfen, Berichtigen, Berechnen, Entscheiden, Kontrollieren.

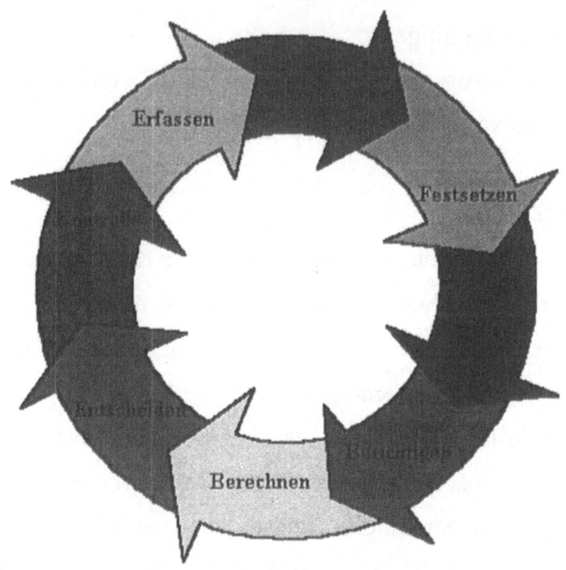

Abbildung 1: Beispiel für typische Aktivitäten bei der Antragsbearbeitung

Der Lösungskern enthält die Komponenten der technologischen Basis und zusätzliche Werkzeuge, die es erlauben, den Kern fachlich zu erweitern und anzupassen. Weiterhin sind für alle Rechtsgebiete Grundfunktionen für gemeinsame fachliche Grundaufgaben enthalten:

- **Leistungskatalog** zur Definition der verschiedenen Leistungsarten der einzelnen Rechtsgebiete.
- **Zahlungsverkehr** zur Definition der Schnittstellen zum Haushalts-, Kassen- und Rechnungswesen.
- **Statistik/Auskünfte** zur Definition der benötigten Statistiken, Auskunftsverfahren und Berichterstattung.
- **Ergonomie** zur Definition der ergonomischen Gestaltung der Software an den DV-Arbeitsplätzen.

- **Zentrale Aufgaben** zur Definition der grundlegenden organisatorischen Vorgaben z.B. Aktenzeichen, Register, generelle Vordrucke, Schnittstellen und Zugriffsrechte.
- **Datenorganisation** zur Definition und Abstimmung z.B. der Synonyme.
- **Verfahrenssicherheit** zur Umsetzung der Anforderungen z.B. an Datenschutz, Datensicherheit und Verarbeitungssicherheit.

Lösungsschalen

Die Softwarearchitektur sieht auf diesen Lösungskern aufsetzende Schalen vor, die eine Konfiguration und Adaption von Varianten des Anwendungssystems ermöglichen. In diesen Schalen sind dann beispielsweise enthalten:

- **Funktionen und Berechnungen zur Einzelfall- und Fall-übergreifenden Bearbeitung** aus den einzelnen Rechtsgebieten.
- **spezielle Statistik und Auskünfte** für Auskunftsverfahren und zur Berichterstattung.
- **organisatorische Konfiguration** als Einbindung in die organisatorischen Strukturen der Dienststellen.
- **spezielle Schnittstellen** zu anderen Verfahren.
- **individuelle Einstellungen** zur Verbesserung der Leistungsfähigkeit, wie z.B. Antwortzeit des zukünftigen Systems.

Bei den Lösungsansätzen fließen die Erfahrungen, die Oracle mit der Entwicklung und Einführung von Standardprodukten im Bereich Oracle Applications hat, in die Konzeption und Methodik ein.

Auswirkungen

Das oben dargestellte Konzept mit einem Lösungskern und seinen Schalen hat direkte Auswirkungen auf die drei genannten Aspekte:

- *Konfiguration,* denn in dem Lösungskern sind Werkzeuge für die aufsetzende Schalen, die eine Konfiguration von Varianten des Anwendungssystems ermöglichen, enthalten.

- *Adaption*, denn der Lösungskern enthält die Komponenten der technologischen Basis und zusätzliche Werkzeuge, die es erlauben, den Kern fachlich zu erweitern und anzupassen.
- *Wiederverwendung*, denn im Lösungskern sind vorzugsweise Softwarebausteine für die Grundaufgaben, -abläufe und -strukturen, die in allen Dienststellen, Kommunen usw. ohne Änderungen genutzt werden können.

Konfiguration

Bei der Anpassung der Softwarelösung an spezifische Besonderheiten der Dienststellen wird unterschieden in

- eine fachliche Konfiguration zur Festlegung des funktionalen Umfangs am jeweiligen Standort und
- eine technische Konfiguration an die technischen Begebenheiten der Hardware und Systemkomponenten.

Fachliche Konfiguration

Die Aufgabe, eine neue Variante für eine Dienststelle zu konfigurieren, besteht darin, mit den entsprechenden Werkzeugen das Verhalten, die Struktur und die Darstellung des Systems anhand von Checklisten auf den spezifischen Umfang anzupassen. Diese Aktivität wird mit *Mapping* bezeichnet, wobei die Grundidee ist, individuelle Profile zu bilden. Zum Beispiel:

- Behördenprofile zur Definition von Briefköpfen, Behördenbezeichnung, Haushaltstiteln, Adressen, ...
- Leistungsprofile zur Selektion der benötigten Rechtsgebiete und Leistungsarten
- Kompetenzprofile zur Zuordnung der Funktionen zu den Dienststellen
- Berechtigungsprofile zur Zuordnung von Anwenderrollen zu Daten
- Durchführungsprofile zur Zuordnung der Berechnungsformel und Parameter

Die Checklisten und die Werkzeuge mit ihren Konfigurationsdialogen werden im Projektverlauf parallel erstellt.

Technische Konfiguration

Aufgabe der technischen Konfiguration ist es, die einzelnen Softwarekonfigurationseinheiten zu einen gesamten Anwendungssystem zu integrieren. Die Softwarearchitektur sieht neben den Arbeitsplätzen für Dialogfunktionen eine Reihe von Services für Kommunikation, Datenbank, Berechnungen, Drucken, Verzeichnis und Auskünften als verteilte Objekte vor.

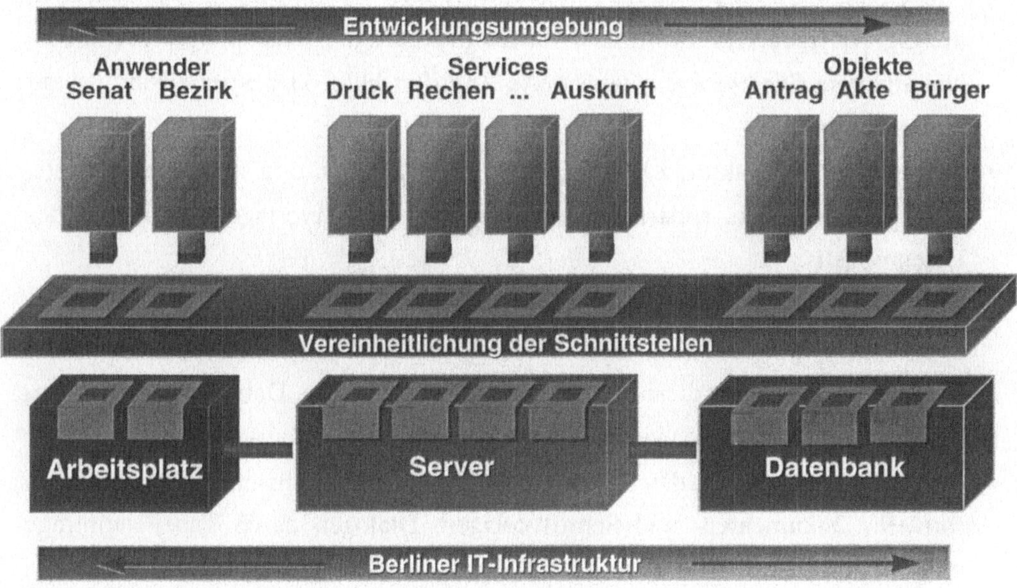

Abbildung 2: Softwarearchitektur

Mit dieser Softwarearchitektur der verteilten Objekte ist es grundsätzlich möglich, das System je nach Hardwareausstattung und Anforderungen an das Datenvolumen und die geforderten Antwortzeiten zu konfigurieren. In einem Extremfall befinden sich alle Client- und Serviceobjekte auf einer Workstation, in dem anderen Extremfall sind diese Objekte vollständig auf unterschiedlicher Hardware verteilt.

Adaption

Während das oben genannte Konzept und die beschriebene Technologie für die Konfiguration in erster Linie die Anpassung von Varianten zum Zeitpunkt der Installation und der Einführung des Anwendungssystems betreffen, hat ein Konzept zur Adaption an Änderungen in den Gesetzen und Vorschriften bedeutende Auswirkungen auf die Softwarepflege und den Betrieb des Systems.

Die Herausforderung liegt darin, ein Problemlösung zu erarbeiten, die im Idealfall später im laufenden Betrieb folgende Anpassungen und Änderungen zuläßt:

- Anpassung des Verhaltens, z.B. an geänderte Rechtsvorschriften, neue Entscheidungsgrundlagen, für bestimmte Zeiträume gültige Berechnungsalgorithmen, überarbeitete Dialogfolgen, veränderte Zugriffsrechte oder Spezialisierung auf Rechtsgebiete.
- Anpassung der Struktur, z.B. zusätzliche Eigenschaften und Werte der Objekte, neue Rechengrößen, andere Dokumenten- und Schriftvorlagen, Änderung des Datenmodells.
- Anpassung der Darstellung, z.B. andere grafische Darstellung.

Die generelle Lösung für diese Anforderungen ist schwierig. Daher wurden die Anforderungen nach ihrem Schwierigkeitsgrad klassifiziert:

- *einfach:* Anpassungen, die durch Konstanten und Wertelisten in Berechnungsformeln, Dokumenten- und Schriftvorlagen, Dialogen durch Tabelleneinträge parametrisiert werden können.
- *einfach bis mittel:* Anpassungen, die durch den Austausch oder Überlagerung von Objekten zum Beispiel durch dynamische Bibliotheken oder Stored Procedures realisiert werden können.
- *einfach bis mittel:* Anpassungen an eine andere grafische Oberfläche, sofern die entsprechende Laufzeitumgebung, zum Beispiel ein Browser, zur Verfügung steht.
- *mittel:* Anpassungen, die durch Änderungen vorhandener Methoden oder Erweiterungen neuer Eigenschaften der Objekte vollzogen werden können.
- *schwierig:* Anpassungen, die nur durch Änderung ganzer Szenarien für Rechtsvorschriften mit neuen Entscheidungsgrundlagen oder beispielsweise einen frei konfigurierbaren Interpreter für gesetzliche Regeln durchgeführt werden können.

Das Konzept eines Lösungskerns mit aufsetzenden Schalen sieht vor, daß die Werkzeuge zur Parametrisierung der Konstanten und Wertelisten, zur Einbindung dynamischer Bibliotheken, Änderung der Objekte und bei Bedarf Interpreter Bestandteil des Lösungskerns sind.

Wiederverwendung

Der Grad der Wiederverwendbarkeit von Softwarebausteinen oder Softwarelösungen hängt von folgenden Voraussetzungen ab:

- es muß ein technologisches Rahmenwerk zur Verfügung stehen, mit dem die wiederverwendeten Bausteine in eine Laufzeitumgebung mühelos eingebunden werden können,
- es muß ein methodisches Rahmenwerk zur Verfügung stehen, damit die Modellierung und der Softwareentwurf von vornherein derart gestaltet wird, daß die Bausteine fachlich für eine Wiederverwendung organisiert und klassifiziert sind,
- es muß eine Entwicklungsumgebung als Bausteinbibliothek zur Verfügung stehen, mit der die Bausteine systematisch abgelegt und wiedergefunden werden können.

Technologisches Rahmenwerk

Als Technologie für die Laufzeitumgebung sieht das Konzept die Oracle Network Computing Architecture (NCA) für verteilte Objekte zur Wiederverwendung sowie leichten Erweiterbarkeit und Wartbarkeit vor.

Eines der technologischen Schlüsselelemente dieser NCA sind Cartridges. Eine Cartridge ist ein von der Network Computing Architecture verwaltetes Objekt, das in einer gängigen Programmiersprache (Java, Visual Basic, C/C++, SQL-Erweiterung etc.) programmiert wurde. Die äußeren Schnittstellen einer Cartridge sind mit IDL (Interface Definition Language) von CORBA definiert und realisiert. So können sie in verteilten Systemen eindeutig identifiziert werden und transparent mit anderen Objekten zusammenarbeiten. Technische Verwaltungsaufgaben werden von den Universal Cartridge Services erledigt:

- Installation
- Registration
- Instantiation

- Invocation
- Administration
- Monitoring
- Security

Neben diesen Services zum Management der Laufzeitumgebung beinhaltet die Architektur Werkzeuge zur Softwareerstellung und -wiederverwendung.

Der Nutzen dieser Grundidee ist in der Abbildung Softwarearchitektur dargestellt: mit dem Ziel der freien Möglichkeit zur Konfiguration und Wiederverwendung ist das System wie weiter oben beschrieben fachlich in Softwarebausteine (Cartridges) gegliedert. Als Bausteine wurden Datenobjekte wie zum Beispiel Akte, Bürger, Antrag oder auch Services wie zum Beispiel Berechnung und Auskunft entworfen. In dieser Architektur sind auch übergreifende Objekte wie zum Beispiel Adreß- und Bankenverzeichnisse enthalten, die dann auch von anderen Verfahren der Verwaltung wiederverwendet - im Sinne von genutzt - werden können. Das Medium, das die verschiedenen Verfahren mit einer einheitlichen Schnittstelle zusammenführt, ist der in der Abbildung in der Mitte dargestellte Objektbus.

Methodisches Rahmenwerk

Um eine praktikable Wiederverwendung der Softwarebausteine zu ermöglichen, ist neben einer technologischen Plattform ein methodische Rahmenwerk dringend notwendig. Die wohl anspruchsvollste Herausforderung in Projekten zur Softwareerstellung ist die Frage, wie die von verschiedenen Mitarbeitern erarbeiten Projektergebnisse klassifiziert, ordentlich abgelegt, wiedergefunden und wiederverwendet werden können. Dazu müssen die Ergebnisse zunächst einmal standardisiert sein. Diese Aufgabe wird durch ein CASE-Werkzeug vereinfacht, jedoch nicht garantiert.

Im beschriebenen Projekt ist die Durchführung nach V-Modell verbindlicher Bestandteil. Das V-Modell regelt die Vorgehensweise für die Softwareentwicklung, beschreibt den Software- Entwicklungsprozeß aus funktionaler Sicht und dient als Checkliste für Software Engineering Standards. Auch die Ideen des V-Modells sind nicht neu oder genial, es gilt eher: "Das V-Modell ist strukturierter, gesunder Menschenverstand". Das V-Modell deckt vollständig den Zyklus der Softwareerstellung ab und wird durchgängig von Werkzeugen und Vorlagen für Projektergebnisse unterstützt.

Besonders sei hier erwähnt, daß sich daher Wiederverwendbarkeit nicht nur auf Softwarebausteine, sondern auch auf die anderen Ergebnisse eines Projektes wie zum Beispiel Konzepte, Anforderungs- und Architekturdokumente bezieht. Dieses gilt

- sowohl für die System-Anforderungsanalyse und den Entwurf, wenn die fachlichen Anforderungen an rechtliche Grundlagen und organisatorische Abläufe in Arbeitsgruppen erfaßt und darauf aufbauend Systemteile (Architektur) entworfen werden,
- als auch für Software-Anforderungsanalyse, wenn die Anforderungen an eine moderne Software präzisiert und durch Prototyping verifiziert werden.

Erst dann werden im Softwareentwurf die Softwarebausteine definiert. Mit den durchgängigen Elementarmethoden der Objektorientierung ist dabei für die Wiederverwendung eine Grundlage geschaffen.

Die Erfahrung aus dem Projekt hat gezeigt, daß es hier in der Praxis des betrieblichen Software Managements noch Restriktionen gibt, da die durchgängige Objektorientierung derzeit noch nicht im V-Modell verankert ist. Die gleichzeitige Forderung nach V-Modell in der derzeitigen Version, Objektorientierung und bewährter Technologie kann nur durch methodische Erweiterungen und Kompromisse aufgelöst werden. Während der Phase Pflichtenheft wurden die klassischen und objektorientierten Ansätze zur Erstellung der Modelle parallel verfolgt. Im Pflichtenheft wurde als erster Schritt unter Nutzung des CASE Werkzeuges Oracle Designer 2000 das Objektmodell aus dem Datenmodell und Funktionsmodell erzeugt. Die Attribute und Beziehungen der Objekte wurden automatisch aus den Entitäten des Datenmodells auf der Basis des Data Dictionary generiert. Die Methoden der Objekte wurden aus dem Funktionsmodell zugeordnet. Damit diese automatische und teilweise manuelle Zuordnung überhaupt möglich wurde, wurden das Daten- und das Funktionsmodell im Sinne des Objektmodells entsprechend strukturiert.

Literatur

[Bai92] Baier, K. u.a.: Konzepte der Software-Wiederverwendung, *Softwaretechnik-Trends,* 12, 4, 40-48, 1992.

[Bus96] Buschmann, F. u.a.: Pattern-oriented Software Architecture, Wiley & Sons, New York, 1996

[Heß92] Heß, H. u.a.: Retrieval wiederverwendbarer Softwarebausteine, *Wirtschaftsinformatik,* 34, 2, 190-200, 1992

[Iso92] Isoda, S.: Experience report on software reuse projects: Its structure, activities, and statistical results, In *14th International Conference on Software Engineering,* Melbourne, Australia, May 11-15, IEEE Computer Society Press, Los Alamitos, California, 320-326, 1992

[Kar95] Karlsson, E.-A.: Software reuse, Wiley & Sons, New York, 1995

[Pre94] Pree, W.: Design Patterns for Object-Oriented Software Development, Addison-Wesley, Reading, Massachusetts, 1994

Retrieval-Dienste für Software-Entwicklungsumgebungen

Andreas Henrich

Abstract

Moderne Software-Entwicklungsumgebungen bestehen aus einer Vielzahl von Werkzeugen, die hinsichtlich der Benutzungsoberfläche, der Ablaufsteuerung und der Datenhaltung integriert sind. Die integrierte Datenhaltung wird dabei i. allg. durch die Verwendung eines Repository unterstützt. Sowohl die kommerziell verfügbaren Repositories als auch die Standards zu diesem Bereich stellen aber leider nur unzureichende Anfragemöglichkeiten zu den verwalteten Daten zur Verfügung. Deshalb haben wir beispielhaft Retrieval-Dienste für den ISO-Standard PCTE entwickelt, die Möglichkeiten des navigierenden Zugriffs mit der Funktionalität einer OQL-artigen Anfragesprache und Techniken des Dokumenten-Retrieval kombinieren. Das vorliegende Papier beschreibt diese Retrieval-Dienste und stellt exemplarisch einige Anwendungen vor.

1. Einleitung

Die Entwicklung komplexer Software-Systeme ist heute ohne entsprechende rechnergestützte Werkzeuge nicht mehr denkbar. Beispiele für derartige Werkzeuge sind Editoren oder Werkzeuge zur Überprüfung von Konsistenzbedingungen. Die Entwicklung geht dabei, wie Nagl [Nag93] es formuliert, *„seit den 80er Jahren von einzelnen Werkzeugen zu größeren Werkzeugansammlungen, von isolierten Werkzeugen zu integrierten Umgebungen, von unspezifischen zu semantischen Werkzeugen"*. Man spricht in diesem Zusammenhang von Software-Entwicklungsumgebungen, die nicht nur eine Ansammlung verschiedener Werkzeuge, sondern eine integrierte Arbeitsumgebung sein wollen.

Um die Integration der Werkzeuge zu unterstützen, wurden *Integrationsrahmen* für Software-Entwicklungsumgebungen definiert. Ein wichtiges Beispiel hierfür ist der ISO- und ECMA-Standard PCTE (Portable Common Tool Environment) [PCT94]. Ein wesentlicher Bestandteil des Integrationsrahmens ist die Datenverwaltungs-Komponente, die auch als *Repository* bezeichnet wird. Im Falle von PCTE handelt es sich dabei um ein operational objektorientiertes

Objekt-Management-System (OMS). Der Standard definiert für dieses OMS zahlreiche Aspekte — wie die DDL, den navigierenden Zugriff auf die Daten, Schema-Operationen, Versionierung, Replikation und Verteilung. Ein deskriptiver, mengenorientierter Zugriff auf die Daten ist aber ebensowenig vorgesehen wie eine inhaltsbasierte Dokumentensuche.

Andererseits gibt es zahlreiche Anwendungen, bei denen ein solcher mengenorientierter Zugriff eine kompakte Definition der benötigten Daten erlauben würde. Beispiele hierfür sind die Suche nach inkonsistenten Objekten, die Aufbereitung von Reports oder die Anwendung von Metriken. Wir haben daher speziell auf die Erfordernisse des dargestellten Anwendungsbereichs zugeschnittene Retrieval-Dienste entwickelt, die den im PCTE-Standard definierten navigierenden Zugriff auf die Daten mit der Option OQL-artiger Anfragen und mit Techniken aus dem Bereich des Dokumenten-Retrieval kombinieren.

Ausgangspunkt unserer Überlegungen ist H-PCTE [Kel92], eine hochperformante, hauptspeicherorientierte Implementierung des PCTE-OMS. Auf dieser Basis haben wir eine an das Datenmodell von PCTE angepaßte OQL-artige Anfragesprache definiert, die Techniken der Dokumentensuche aus dem Information-Retrieval integriert. Diese P-OQL genannte Anfragesprache werden wir in Abschnitt 2. beschreiben. Einen weiteren Schwerpunkt unserer Arbeiten, den wir in Abschnitt 2.6. kurz darstellen werden, bildet die effiziente Implementierung dieser Anfragesprache auf Basis entsprechender Zugriffsstrukturen. Schließlich werden wir in Abschnitt 3. exemplarisch einige Werkzeuge vorstellen, die mit Hilfe der beschriebenen Anfragesprache realisiert worden sind. Abbildung 1 faßt den Aufbau einer Software-Entwicklungsumgebung (SEU) im betrachteten Szenario zusammen und verdeutlicht die Einordnung der in den folgenden Abschnitten beschriebenen Teilaspekte.

An dieser Stelle soll noch kurz auf einige vergleichbare Ansätze hingewiesen werden. Ein erster Vorschlag im Hinblick auf eine Anfragesprache für System-Entwicklungsumgebungen ist DQMCS [TTB+90]. DQMCS kann als wiederverwendbarer Baustein für Werkzeugentwickler verstanden werden, der zur Implementierung von Anfragediensten verwendet werden kann. In [Kel93a] werden einige Anforderungen an Retrieval-Dienste skizziert. Dabei wird insbesondere die Notwendigkeit von Dokumenten-Retrieval-Funktionalitäten hervorgehoben. Zwei einfache Anfragesprachen für PCTE werden in [Bir95] beschrieben. Während eine dieser Sprachen Navigationsbefehle um Ausgabedefinitionen ergänzt, hat die andere eine SQL-artige Syntax. Sie orientiert sich

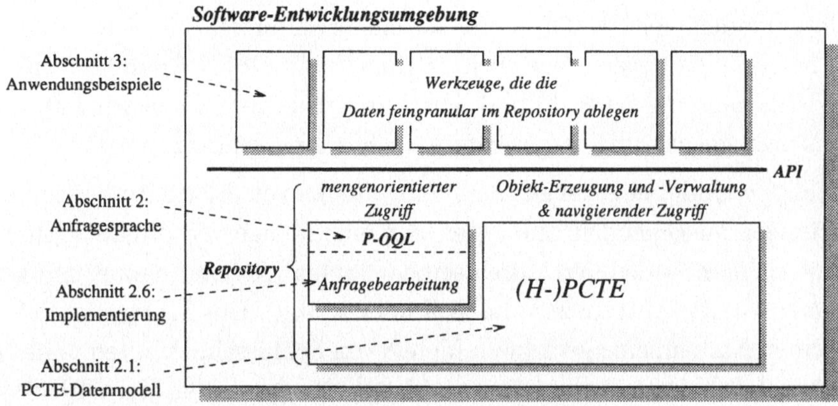

Abbildung 1: Aufbau einer SEU im betrachteten Szenario

aber trotzdem stark an navigierenden Operationen.

Während die obigen, auf System-Entwicklungsumgebungen ausgerichteten Ansätze den Dokumenten-Retrieval-Aspekt vernachlässigen, gibt es zahlreiche Ansätze zur Kombination mengenorientierter Anfragesprachen mit Dokumenten-Retrieval-Funktionalitäten im Umfeld von SGML, WWW oder Hypertextsystemen (vgl. hierzu z.B. [KS95, CACS94, CGMB95, VAB96]). Diese Ansätze gehen aber i. allg. von schwach strukturierten Daten aus, während wir auf dem ausdrucksstarken Datenmodell von PCTE aufsetzen. Daneben gibt es auch im Bereich des Information-Retrieval einige Ansätze zur Integration von Datenbanken und Dokumenten-Retrieval (vgl. z.B. [Fuh93]) und zur Unterstützung von Software-Bibliotheken (vgl. z.B. [MBK91] oder [Hen91]), die hier aber aus Platzgründen nicht im Detail betrachtet werden können.

2. Die Anfragesprache

Als Ausgangsbasis für die P-OQL[1] (PCTE Object Query Language) genannte Anfragesprache wurde OQL [Cat93] gewählt. Für den beabsichtigten Anwendungsbereich mußte OQL aber zum einen an das Datenmodell von PCTE angepaßt und zum anderen um Möglichkeiten des Dokumenten-Retrieval ergänzt werden. Im folgenden beschreiben wir daher kurz das Datenmodell von PCTE bevor wir auf die Besonderheiten von P-OQL eingehen.

[1] Eine ausführliche Darstellung einer ersten Version von P-OQL findet sich in [Hen95].

2.1. Das Datenmodell von PCTE

Das Datenmodell von PCTE kann als ein erweitertes Entity-Relationship-Modell beschrieben werden. Die Objektbank enthält Objekte und Beziehungen. Die Beziehungen sind i.d.R. bidirektional, d.h. jede Beziehung wird durch ein Paar aus zwei gegenläufigen, gerichteten *Links* repräsentiert.

Ein *Objekttyp* ist dabei gegeben durch eine Menge von Attributen, eine Menge von zulässigen ausgehenden Linktypen und eine Menge von direkten Elterntypen. Ein *Attribut* ist definiert durch seinen Namen (Attributname) und seinen Wertebereich (Attributtyp; z.B. *natural* oder *string*). Ein *Linktyp* ist definiert durch seinen Namen, eine geordnete Menge von Schlüsselattributen, eine Menge von Nicht-Schlüsselattributen, eine Menge der erlaubten Ziel-Objekttypen und eine Kategorie, die gewisse semantische Eigenschaften des Links festlegt. Beispielsweise wird für Links der Kategorie *reference* die referentielle Integrität (also die Existenz des Zielobjekts) garantiert und bei Links der Kategorie *composition* gilt das Zielobjekt als Komponente des Ausgangsobjekts.

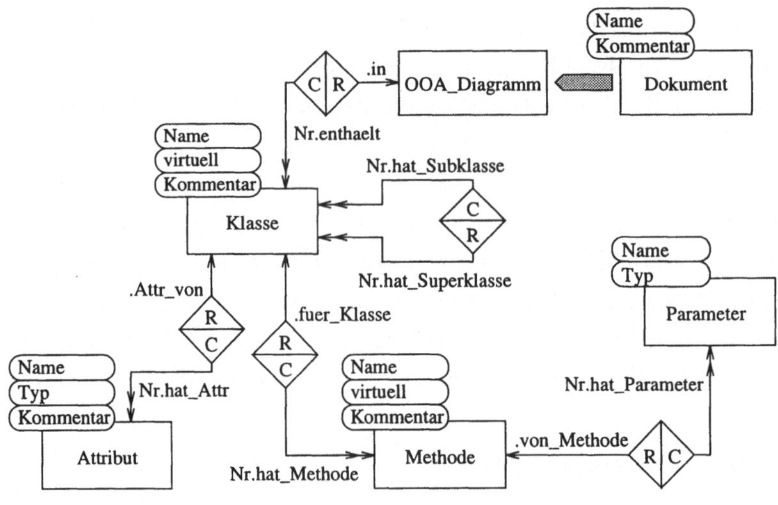

Abbildung 2: Beispiel-Schema für OOA-Diagramme

Abbildung 2 beschreibt ein einfaches PCTE Schema für OOA-Diagramme. Objekttypen sind durch Rechtecke und die zugehörigen Attribute durch Ovale repräsentiert. Die Linktypen, die die Beziehungen zwischen den Objekten repräsentieren, sind durch Pfeile dargestellt. Eine doppelte Pfeilspitze zeigt an, daß der Link die Kardinalität *viele* hat. Links der Kardinalität *viele* müssen in PCTE über ein Schlüsselattribut verfügen. In unserem Beispiel wird für diesen Zweck immer das numerische Attribut *Nr* verwendet. Ein *C* oder *R* in den Dreiecken in der Mitte der Linie, die ein Paar von Links darstellt, zeigt dabei an, daß der Link in der entsprechenden Richtung die Kategorie *composition* oder *reference* hat. Schließlich enthält das Schema eine Vererbungsbeziehung zwischen den Objekttypen *Dokument* und *OOA_Diagramm*.

2.2. Die Darstellung des Ergebnisses

Eine wichtige Anforderung an Anfragesprachen ist die Forderung nach Abgeschlossenheit über dem jeweiligen Datenmodell. Da PCTE aber keine Mengen oder Listen kennt, haben wir zur Darstellung des Ergebnisses einer P-OQL-Anfrage einen abstrakten Datentyp (ADT) Value eingeführt, der einen PCTE Attributwert, eine Objektreferenz[2], einen Linkdeskriptor oder eine Menge, Multimenge, Liste oder Struktur von Values aufnehmen kann. Durch die Integration von Objektreferenzen und Linkdeskriptoren in diesen ADT bleiben wesentliche Vorteile einer abgeschlossenen Anfragesprache erhalten.

2.3. Berücksichtigung der Links

Im Verhältnis zum Datenmodell der ODMG, auf dem OQL basiert, muß eine an PCTE angepaßte Anfragesprache die ausdrucksstärkeren Beziehungen (Links) reflektieren. In P-OQL können deshalb nicht nur Objekttypen, sondern auch Linktypen zur Definition der Basismengen einer Anfrage verwendet werden.

So bestimmt die Anfrage '`select ^Name, /Name from hat_Subklasse`' alle Vererbungsbeziehungen zwischen Klassen. Mit dieser Anfrage wird zu jedem Link des Typs *hat_Subklasse* der Name der Superklasse und der Name der Subklasse bestimmt (durch ein vorangestelltes '^' bzw. '/' bezieht man sich auf das Ausgangs- bzw. Zielobjekt eines Links).

Eine weitere Berücksichtigung der Links erfolgt durch die Integration mächtiger regulärer Pfadausdrücke[3] in die Sprache. Damit können von einem Ob-

[2] Objektreferenzen und Linkdeskriptoren werden in PCTE zum navigierenden Zugriff auf Objekte und Links verwendet und mit Hilfe navigierender Operationen gesetzt.

[3] P-OQL orientiert sich hier zum Teil an den in der Literatur präsentierten Ansätzen zu Pfadausdrücken (vgl. z.B. [BV93, Sci94, FLU94]). Während in diesen Papieren Pfadausdrücke aber oftmals als eine Art Abkürzung verwendet werden, um eine gesuchte Beziehung zu definieren, definieren Pfadausdrücke in P-OQL eine Multimenge von Values.

jekt oder Link aus entfernte Informationen angesprochen werden. Ein Beispiel hierfür ergibt sich, wenn überprüft werden soll, ob die in der Objektbank abgelegten Vererbungsbeziehungen zyklenfrei sind. Die folgende Anfrage sucht nach den Namen aller Klassen, die in einem Zyklus enthalten sind:

```
select Name from Klasse where . in [_.hat_Subklasse]+/->.
```

Durch den Punkt in der `where`-Klausel wird die aktuelle Klasse (d.h. das entsprechende Objekt) adressiert. Der Pfadausdruck '`[_.hat_Subklasse]+/->.`' bestimmt die Menge der direkten und indirekten Subklassen der aktuellen Klasse. '`_.hat_Subklasse`' bedeutet dabei, daß Links des Typs *hat_Subklasse* mit beliebigen Schlüsselattributwerten verfolgt werden sollen. Durch den Einschluß dieser Linkdefinition in das Iterationskonstrukt '`[...]+`' wird definiert, daß Pfade verfolgt werden sollen, die aus einem oder mehreren Links bestehen, die der Linkdefinition entsprechen. Dem Iterationskonstrukt liegt dabei eine entsprechende Fixpunktsemantik zugrunde.

2.4. Integration mit dem navigierenden Zugriff

Um Objektreferenzen und Linkdeskriptoren, die durch navigierende PCTE-Operationen bestimmt worden sind, als Ausgangspunkt einer P-OQL-Anfrage zu verwenden, können *Sequenzen* (ein von PCTE zur Verwaltung von Mengen definierter ADT) mit Objektreferenzen oder Linkdeskriptoren über die Programmierschnittstelle von P-OQL übergeben und in Anfragen als Basismengen verwendet werden. Ferner wird immer dann, wenn in der `select`-Klausel einer Anfrage durch einen Punkt ein Objekt (bzw. Link) adressiert wird, eine entsprechende Objektreferenz (bzw. ein entsprechender Linkdeskriptor) in das Ergebnis der Anfrage aufgenommen, die als Ausgangspunkt für navigierende PCTE-Operationen dienen kann. So liefert die Anfrage '`select Name, . from Klasse`' zu jeder Klasse den Namen und eine Objektreferenz.

2.5. Möglichkeiten des Dokumenten-Retrieval

Die in einem Repository verwalteten Objekte haben oft den Charakter von Dokumenten mit einem relativ hohen Textanteil – z.B. in Form von als Freitext abgelegten Kommentaren. Zur Suche nach solchen Dokumenten bieten sich Techniken des Dokumenten-Retrieval an. Dabei ist nicht nur an die reine Zeichenkettensuche (analog zum UNIX-Befehl `grep`) zu denken, sondern z.B. auch an die Suche mit Hilfe des Vektorraummodells [SWY75]. In diesem

Modell werden Dokumente und Informationswünsche durch Beschreibungsvektoren dargestellt, auf denen ein Ähnlichkeitsmaß definiert ist. Den Ausgangspunkt bildet ein Vokabular, das die für potentielle Informationswünsche relevanten Begriffe enthält [4]. Zu den einzelnen Dokumenten wird dann die Relevanz bezüglich der Begriffe aus dem Vokabular bestimmt. Diese Relevanzwerte werden in einen Beschreibungsvektor für das Dokument eingetragen, der für jeden Begriff im Vokabular eine Komponente enthält. Man kann dann zu einem durch einen Anfragetext definierten Informationswunsch ebenfalls einen Beschreibungsvektor erzeugen, und Dokumente mit einem ähnlichen Beschreibungsvektor im Datenbestand suchen.

Wir haben in P-OQL beide Arten der Suche nach Dokumenten integriert. Für die zeichenkettenbasierte Suche enthält P-OQL einen an das UNIX Kommando *grep* angelehnten binären Operator, der als ersten Operanden einen regulären Ausdruck und als zweiten Operanden eine Dokumentdefinition erwartet. Dieser Operator erzeugt eine Menge, die für jede Zeichenkette innerhalb des Dokumentes, die dem regulären Ausdruck entspricht, einen Eintrag enthält. Als Dokumentdefinition können dabei nicht nur String-Attribute, sondern auch Objekte, Links sowie Mengen, Multimengen, Listen oder Strukturen verwendet werden. Der *grep* Operator bezieht sich dabei immer auf die textuellen Anteile der Kollektionen bzw. auf die String-Attribute der Objekte und Links.

Zur Ähnlichkeitssuche nach dem Vektorraummodell enthält P-OQL einen Operator `D_vector`, der zu einem analog zum *grep* Operator definierten Dokument einen Beschreibungsvektor erzeugt. Die Ähnlichkeit zwischen zwei Beschreibungsvektoren kann dann mit Hilfe des `sim`-Operators bestimmt werden.

Gesucht seien z.B. Klassendefinitionen, die zu einem textuell beschriebenen Informationswunsch relevant sind:

```
head[25]
  sort-
    (select (D_vector "<als Text formulierter Informationswunsch>"
                sim
                D_vector (., _.hat_Methode/->., _.hat_Attr/->.),

          from Klasse)
```

Die durch '`(., _.hat_Methode/->., _.hat_Attr/->.)`' definierten Dokumente bestehen dabei aus der Klasse selbst sowie ihren Methoden- und At-

[4]Techniken zur automatischen Erzeugung eines solchen Vokabulars für eine gegebene Dokumentmenge werden zum Beispiel in [Cro90] diskutiert.

tributdefinitionen. Die Select-Anweisung liefert als Ergebnis Paare aus einem Ähnlichkeitswert in Relation zum gegebenen Anfragetext und einer Objektreferenz für die Klasse. Durch `sort-` werden diese Paare nach der Ähnlichkeit absteigend sortiert. Der Befehl `head[25]` reduziert das Ergebnis schließlich auf die 25 „ähnlichsten“ Klassen (vgl. hierzu auch [Hen96b]).

Um das von PCTE verwendete Konzept der Linkkategorien auch bei der Definition der Komponenten, die ein Dokument bilden sollen, ausschöpfen zu können, kann man sich innerhalb eines regulären Pfadausdrucks auch auf die Linkkategorien beziehen. Wollten wir z.B. bei der Berechnung der Beschreibungsvektoren für die Klassen jeweils alle über Links der Kategorie *composition* erreichbaren Objekte mit Ausnahme der Subklassen betrachten, so könnten wir dies durch '`D_vector ([{c shield hat_Subklasse}]*/->.)`' ausdrücken. Das Konstrukt '`{c shield hat_Subklasse}`' adressiert dabei alle Links der Kategorie *composition* mit Ausnahme der Links des Typs *hat_Subklasse*. Durch den Einschluß dieser Linkdefinition in das Iterationskonstrukt '`[...]*`' wird definiert, daß Pfade verfolgt werden sollen, die aus keinem, einem oder mehreren Links bestehen, die der Linkdefinition entsprechen.

2.6. Zugriffsstrukturen und Anfragebearbeitung

Da die beschriebene Anfragesprache in wesentlichen Punkten — mächtige reguläre Pfadausdrücke, Integration von Dokumenten-Retrieval-Funktionalität — über die von herkömmlichen Anfragesprachen gebotene Funktionalität hinausgeht, konnte zu ihrer effizienten Implementierung nur begrenzt auf die im Datenbankbereich bekannten Konzepte zurückgegriffen werden.

Durch die Integration des Vektorraummodells in die Anfragesprache und die damit mögliche Kombination einer Ähnlichkeitssuche mit Bedingungen über Standardattributen erscheint im betrachteten Umfeld der Einsatz mehrdimensionaler Zugriffsstrukturen vielversprechend. Die Idee ist, daß einzelne Dimensionen der Zugriffsstruktur „Standardattributen“ wie z.B. einem Namen oder der Anzahl ausgehender Links eines bestimmten Typs entsprechen. Eine weitere logische Dimension kann dann durch einen Beschreibungsvektor definiert sein. Eine Indexdefinition könnte damit z.B. wie folgt aussehen:

```
Index for Klasse:
   1. Name
   2. count _.hat_Methode->.
   3. D_vector (., _.hat_Attr/->., _.hat_Methode/->.)
```

Die erste Dimension entspricht dabei dem Namen der Klasse, die zweite der Anzahl der Methoden und die dritte dem Beschreibungsvektor für das durch den gegebenen Pfadausdruck definierte Dokument (physisch handelt es sich dabei um die Dimensionen 3 bis $3+t$ wenn das Vokabular t Begriffe enthält).

Durch eine Kombination der in [HM95] für Multiattribut-Zugriffsstrukturen beschriebenen Techniken mit den in [Hen96a] dargestellten Techniken zur Bearbeitung von Ähnlichkeitssuchen im Vektorraum können mit einer solchen Zugriffsstruktur Anfragen effizient bearbeitet werden, die eine Ähnlichkeitssuche im Vektorraum mit Bedingungen über Standardattributen kombinieren. Dabei wird die Ähnlichkeitssuche dadurch beschleunigt, daß Teile der Zugriffsstruktur, für die die Bedingungen über den Standardattributen nicht erfüllt sein können, aus der Betrachtung ausgeschlossen werden.

3. Anwendungsbeispiele

Im folgenden wollen wir exemplarisch vier Werkzeuge vorstellen, die mit Hilfe der beschriebenen Retrieval-Dienste implementiert wurden.

3.1. Report-Generator

Im Rahmen eines Software-Entwicklungsprojekts ist es häufig erforderlich, Informationen in Form von Reports aufzubereiten. Die Informationen für einen solchen Report lassen sich i. allg. relativ leicht mit Hilfe einer P-OQL-Anfrage ermitteln. Wir haben daher einen Report-Generator entwickelt, der dem Anwender Reports anbietet, die durch eine Kombination aus einer Anfrage und einer Layout-Beschreibung definiert werden.

Abbildung 3 zeigt exemplarisch eine solche Report-Definition. Die erste Zeile definiert den Namen des Reports und den adressierten Objekttyp (*OOA_Diagramm*). Ferner ist hier spezifiziert, daß der Name des OOA-Diagramms, für das der Report erstellt werden soll, als Parameter in die Report-Definition eingesetzt werden muß. D.h. der konkrete Name wird immer dann, wenn ein Report des Typs *„Klassenaufstellung“* angefordert wird, vom Report-Generator erfragt und in der Report-Definition überall dort ersetzt, wo '`#$1#`' steht. Den nächsten Teil der Report-Definition bildet die P-OQL-Anfrage, mit der die Informationen für den Report gesucht werden. Im Beispiel handelt es sich um eine sortierte Liste der Klassen mit ihren Attributen. Die Layout-Beschreibung

```
Report "Klassenaufstellung" for OOA_Diagramm, Parameter: Name
Query:  sort+ (select K:Name, K:_.hat_Attr/->Name
                 from D in OOA_Diagramm, K in (D:_.enthaelt/->.)
                 where D:Name = "#$1#")
Layout(LaTeX):  \documentstyle{article}
                \begin{document}
                #BEGIN_ITERATE .#
                  \newpage
                  {\bf Klasse: #.e.1.#}

                  {\em Attribute:}
                  \begin{itemize}
                  #BEGIN_ITERATE .e.2.#
                    \item #.e.2.e.#
                  #END_ITERATE .e.2.#
                  \end{itemize}
                #END_ITERATE .#
                \end{document}
```

Abbildung 3: Beispiel für eine Report-Definition

schließt die Report-Definition ab. Im Beispiel verwenden wir hierzu die LaTeX-Syntax. Die Layout-Beschreibung enthält in '#'-Zeichen eingeschlossene Steueranweisungen, die definieren, wie das Ergebnis der Anfrage in den Report eingearbeitet werden soll. Soll eine bestimmte Folge von Formatierungsanweisungen z.B. für jedes Element einer Liste wiederholt werden, so sind die entsprechenden Formatierungsanweisungen mittels '#BEGIN_ITERATE <Value-*Zugriff*>#' und '#END_ITERATE <Value-*Zugriff*>#' zu klammern. Auf die einzelnen Komponenten des Anfrageergebnisses wird durch Komponentendefinitionen zugegriffen. So greift man mit 'e.' auf ein Element einer Menge, Multimenge oder Liste zu. Mit '2.' wird auf die zweite Komponente einer Struktur zugegriffen.

3.2. Schätzung der Software-Entwicklungskosten

Zur Schätzung des Aufwands, der für eine geplante Software-Entwicklung erforderlich sein wird, sind zahlreiche Verfahren vorgestellt worden. Besonders auf frühen Analyseergebnissen basierende Verfahren, wie die *Object-Point-Methode* [Sne96], bieten sich dabei für eine Repository-basierte Implementierung an. Bei der Object-Point-Methode wird die Schätzung in drei Teilschritten durchgeführt, in denen je eine Kenngröße für den *Codieraufwand*, den *Aufwand für die Klassenintegration und den Integrationstest* und den *Aufwand für den Systemtest* bestimmt wird. Insgesamt ergeben sich folgende Formeln:

$$\begin{aligned}
\text{Class-Points} &= (\text{Attribute} + \text{Relationen} \times 2 + \text{Methoden} \times 3) \times \text{Neuheit} \\
\text{Message-Points} &= (\text{Parameter} + \text{Quellen} \times 2 + \text{Ziele} \times 2) \times \text{Komplexität} \times \text{Neuheit} \\
\text{Process-Points} &= (\text{Prozeßtyp} + \text{Varianten}) \times \text{Komplexität} \\
\text{Object-Points} &= \text{Class-Points} + \text{Message-Points} + \text{Process-Points}
\end{aligned}$$

Die so gewonnene Ausgangsschätzung wird dann noch durch Korrekturfaktoren adjustiert, die Qualitätsanforderungen (QF) — wie Zuverlässigkeit oder Portabilität — und Projekt-Einflußfaktoren (EF) — wie die eingesetzten Methoden — berücksichtigen:

$$\text{Adjusted-Object-Points} = \text{Object-Points} \times \text{QF} \times \text{EF}$$

Ein Adjusted-Object-Point entspricht dann laut Sneed ungefähr einem Entwicklungsaufwand von 0,25 Personentagen.

Wenn nun entsprechende Dokumente zur Klassendefinition, zum Nachrichtenaustausch und zur Prozeßmodellierung im Repository abgelegt sind, dann können die Object-Points direkt mit einer Anfrage aus dem Repository gewonnen werden. Abbildung 4 zeigt eine solche Anfrage, die ein entsprechendes Schema unterstellt und zur Umwandlung von Aufzählungstypen in numerische Werte auf die in P-OQL verfügbare `case`-Anweisung zurückgreift. Diese Anfrage bestimmt die Object-Points für ein Projekt namens *„Abrechnung“*.

Zur Berechnung der Class-Points, Message-Points und Process-Points wird dabei jeweils nach dem gleichen Muster vorgegangen, das wir für die Message-Points kurz verdeutlichen wollen. Ausgangspunkt der Anfrage sind die Projekte (Basismenge `P`). In der `where`-Klausel wird aus diesen Projekten das Projekt mit Namen *„Abrechnung“* selektiert. In der zweiten Basismenge `N` werden alle vom aktuellen Projekt aus erreichbaren Komponenten adressiert. Diese werden in der `where`-Klausel auf Nachrichten eingeschränkt. Für jede Nachricht wird mit Hilfe des `count`-Operators die Anzahl der ausgehenden Links der Typen *hat_Parameter*, *Quelle* und *Ziel* ermittelt.

Die Anfrage aus Abbildung 4 bildet auch die Grundlage für das von uns entwickelte Werkzeug zur Aufwandsschätzung nach der Object-Point-Methode (vgl. hierzu auch [Hen97]). Dieses Werkzeug geht insgesamt wie folgt vor: Zunächst wird der Name des zu betrachtenden Projekts erfragt. Die Berechnung der Object-Points erfolgt dann analog zu der in Abbildung 4 gegebenen

```
(sum                                          // summiere die Class-Points
   (select ((count K:_.hat_Attr->.
            + ((2 * count K:_.Relation->.)
               + (3 * count K:_.hat_Methode->.))) * K:Neuheit)
      from P in Projekt, K in (P:[{c}]+/->.)
      where P:Name = "Abrechnung" and K:. is of type Klasse)
 + (sum                                       // summiere die Message-Points
      (select ((count N:_.hat_Parameter->.
               + ((2 * count N:_.Quelle->.)
                  + (2 * count N:_.Ziel->.)))
               * (N:Neuheit
                  * (case N:Komplexitaet = NIEDRIG ==> +0.75,
                          N:Komplexitaet = MITTEL  ==> +1.00,
                          N:Komplexitaet = HOCH    ==> +1.25)))
         from P in Projekt, N in (P:[{c}]+/->.)
         where P:Name = "Abrechnung" and N:. is of type Nachricht)
  + sum                                       // summiere die Process-Points
      (select (((case Pz:Prozesstyp = SYSTEM   ==> 6,
                      Pz:Prozesstyp = BATCH    ==> 2,
                      Pz:Prozesstyp = ONLINE   ==> 4
                      Pz:Prozesstyp = REALTIME ==> 8) + Pz:Varianten)
               * (case Pz:Komplexitaet = NIEDRIG ==> +0.75,
                       Pz:Komplexitaet = MITTEL  ==> +1.00,
                       Pz:Komplexitaet = HOCH    ==> +1.25))
         from P in Projekt, Pz in (P:[{c}]+/->.)
         where P:Name = "Abrechnung" and Pz:. is of type Prozess)))
```

Abbildung 4: P-OQL-Anfrage zur Bestimmung der Object-Points

Anfrage. Das Ergebnis der Anfrage wird in einem entsprechenden Fenster dargestellt. Der Benutzer kann dann die Korrekturfaktoren angeben. Im Anschluß werden die Adjusted-Object-Points berechnet und angezeigt.

3.3. Dokumenten-Retrieval

Um die Dokumenten-Retrieval-Funktionalität von P-OQL auch dem Nutzer der Software-Entwicklungsumgebung zugänglich zu machen, haben wir ein einfaches Frontend entworfen, mit dem *„Dokumente"* inhaltsbasiert oder zeichenkettenorientiert gesucht werden können. Abbildung 5 zeigt dieses Werkzeug.

Im ersten Teilfenster des Suchwerkzeugs (*Addressed Entity*) kann der Benutzer den als *„Dokumenttyp"* anzusprechenden Objekttyp auswählen. Er kann dabei

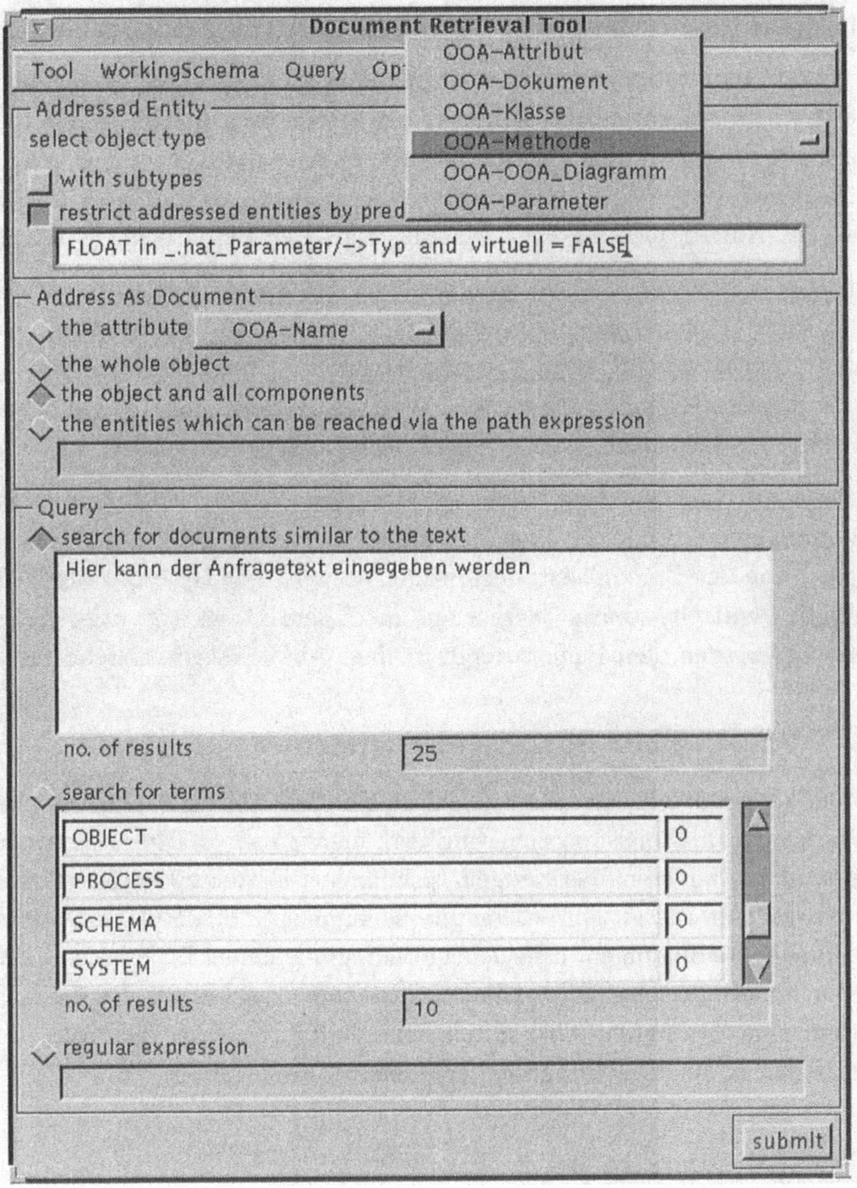

Abbildung 5: Ein Werkzeug zur Suche nach *„Dokumenten“*

angeben, ob er nur an Objekten des Typs selbst oder auch an Subtypen interessiert ist. Zusätzlich kann hier ggf. noch ein einschränkendes Prädikat angegeben werden. Im Beispiel wird dieses Prädikat genutzt, um die Suche auf nichtvirtuelle Methoden mit mindestens einem FLOAT-Attribut einzuschränken.

Im zweiten Teilfenster (*Address as Document*) ist anzugeben, ob für ein konkretes Objekt des gewählten Typs (1) ein bestimmtes Attribut oder (2) das Objekt selbst oder (3) das Objekt und alle über *composition*-Links erreichbaren Teilobjekte als Dokument betrachtet werden soll. Ferner ist hier für den erfahrenen Nutzer auch die Eingabe eines regulären Pfadausdrucks möglich.

Im dritten Teilfenster (*Query*) ist schließlich das Suchkriterium zu definieren. Hierzu kann (1) ein Anfragetext angegeben werden, zu dem ähnliche *„Dokumente"* gesucht werden, oder (2) eine Anzahl von Begriffen aus dem Vokabular mit entsprechenden Gewichten versehen werden oder (3) ein regulärer Ausdruck zur zeichenkettenorientierten Suche angegeben werden.

Das Ergebnis einer auf diese Weise spezifizierten Anfrage wird dann zunächst als Liste in einem Standard-Ergebnis-Viewer angezeigt. Wird ein *„Dokument"* (Objekt) aus der Ergebnisliste angewählt, so wird mit Hilfe der Meta-Werkzeugbank ermittelt, welche Werkzeuge zu diesem Objekttyp existieren. Die Werkzeuge werden dem Benutzer ggf. in einer Auswahlleiste angeboten.

3.4. Werkzeug zur Konsistenzsicherung

Bei dem letzten Werkzeug, das wir hier ansprechen wollen, handelt es sich um ein Werkzeug zur Konsistenzsicherung (vgl. hierzu auch [HD96]). Dieses Werkzeug kann aus anderen Werkzeugen (primär aus Editoren) gestartet werden. Es überprüft für das aktuell bearbeitete Dokument (z.B. ein OOA-Diagramm) vordefinierte Bedingungen. Eine solche Bedingung kann z.B. festlegen, daß die Klassen innerhalb eines OOA-Diagramms eindeutige Namen haben müssen. Jede einzelne Bedingung wird durch eine P-OQL Anfrage definiert, die die Objekte sucht, die die Bedingung verletzen.

4. Projektstatus und Ausblick

Die in diesem Papier vorgestellte Funktionalität der Retrieval-Dienste (incl. der Zugriffsstrukturen) ist auf H-PCTE realisiert worden und in weiten Teilen über FTP (`ftp://ftp.informatik.uni-siegen.de/pub/pi/hpcte/`) verfügbar.

Damit existiert eine fundierte Ausgangsbasis für weiterführende Arbeiten in dem beschriebenen Bereich. Gedacht ist dabei derzeit insbesondere (1) an die Weiterentwicklung der Dokumenten-Retrieval-Funktionalität im Hinblick auf nichttextuelle Aspekte der Ähnlichkeit, (2) an die Entwicklung eines umfassenden Werkzeugs zur Unterstützung des Projektmanagements, (3) an ein Werkzeug zur Suche nach wiederverwendbaren Komponenten, sowie (4) an die Weiterentwicklung der Zugriffsstrukturen und Optimierungstechniken.

Literatur

[Bir95] B. Bird. An open systems SEE query language. In *Proc. 7th Conf. on Software Engineering Environments*, S. 34–47, Noordwijkerhout, 1995.

[BV93] J. Van den Bussche und G. Vossen. An extension of path expressions to simplify navigation in object-oriented queries. In *Deductive and Object-Oriented Databases*, volume 760 of *LNiCS*, S. 267–282, Phoenix, 1993.

[CACS94] V. Christophides, S. Abiteboul, S. Cluet und M. Scholl. From structured documents to novel query facilities. In *Proc. ACM SIGMOD Intl. Conf. on Management of Data*, S. 313–324, Minneapolis, 1994.

[Cat93] R. Cattell, Hrsg. *The Object Database Standard: ODMG-93.* Morgan Kaufmann, San Mateo, 1993.

[CGMB95] C. Clifton, H. Garcia-Molina und D. Bloom. Hyperfile: A data and query model for documents. *VLDB Journal*, 4(1):45–86, 1995.

[Cro90] J.C. Crouch. An approach to the automatic construction of global thesauri. *Information Processing and Management*, 26(5):629–640, 1990.

[FLU94] J. Frohn, G. Lausen und H. Uphoff. Access to objects by path expressions and rules. In *Proc. 20th Intl. Conf. on VLDB*, S. 273–284, 1994.

[Fuh93] N. Fuhr. A probabilistic relational model for the integration of IR and databases. In *Proc. 16th Annual Intl. ACM SIGIR Conf.*, S. 309–317, Pittsburgh, 1993.

[HD96] A. Henrich und D. Däberitz. Using a query language to state consistency constraints for repositories. In *Proc. 7th Intl. Conf. on Database and Expert Systems Applications*, volume 1134 of *LNiCS*, S. 59–68, Zürich, 1996.

[Hen91] S. Henninger. Retrieving software objects in an example-based programming environment. In *Proc. 14th Annual Intl. ACM SIGIR Conf.*, S. 251–260, Chicago, 1991.

[Hen95] A. Henrich. P-OQL: an OQL-Oriented Query Language for PCTE. In *Proc. 7th Conf. on Software Engineering Environments*, S. 48–60, Noordwijkerhout, 1995.

[Hen96a] A. Henrich. Adapting a spatial access structure for document representations in vector space. In *Proc. 5th Intl. Conf. on Information and Knowledge Management*, S. 19–26, Rockville, 1996.

[Hen96b] A. Henrich. Document retrieval facilities for repository-based system development environments. In *Proc. 19th Annual Intl. ACM SIGIR Conf.*, S. 101–109, Zürich, 1996.

[Hen97] A. Henrich. Repository Based Software Cost Estimation. In *Proc. 8th Intl. Conf. on Database and Expert Systems Applications, LNiCS*, Toulouse, September 1997.

[HM95] A. Henrich und J. Möller. Die Nutzung mehrdimensionaler Zugriffsstrukturen für Anfragen über Standardattributen. In *Tagungsband zur GI-Fachtagung „Datenbanksysteme für Büro, Technik und Wissenschaft"*, Informatik Aktuell, S. 212–231, Dresden, 1995. Springer.

[Kel92] U. Kelter. H-PCTE: A high-performance object management system for system development environments. In *Proc. 16th Annual Intl. Computer Software and Applications Conf.*, S. 45–50, Chicago, 1992.

[Kel93a] U. Kelter. An information retrieval common service based on H-PCTE. In *Proc. 6th Conf. on Software Engineering Environments*, S. 101–108, Reading, UK, 1993.

[KS95] D. Konopnicki und O. Shmueli. W3QS: A query system for the world-wide web. In *Proc. 21th Intl. Conf. on VLDB*, S. 54–65, Zürich, 1995.

[MBK91] Y.S. Maarek, D.M. Berry und G.E. Kaiser. An information retrieval approach for automatically constructing software libraries. *IEEE, Transactions on Software Engineering*, 17(8):800–813, 1991.

[Nag93] M. Nagl. Software-Entwicklungsumgebungen: Einordnung und zukünftige Entwicklungslinien. *Informatik-Spektrum*, 16(5):273–280, 1993.

[PCT94] Portable Common Tool Environment - Abstract Specification / C Bindings. ISO IS 13719-1/-2, 1994 sowie ECMA-149/-158, 1993.

[Sci94] E. Sciore. Query abbreviation in the entity-relationship data model. *Information Systems*, 19(6):491–511, 1994.

[Sne96] H.M. Sneed. Schätzung der Entwicklungskosten von objektorientierter Software. *Informatik-Spektrum*, 19(3):133–140, 1996.

[SWY75] G. Salton, A. Wong und C.S. Yang. A vector space model for automatic indexing. *Communications of the ACM*, 18(11):613–620, 1975.

[TTB+90] M. Tedjini, I. Thomas, G. Benoliel, F. Gallo und R. Minot. A query service for a software engineering database system. In *Proc. 4th ACM Symp. on Software Development Environments*, ACM SIGSOFT Newsletter 15:6, S. 238–248, 1990.

[VAB96] M. Volz, K. Aberer und K. Böhm. Applying a flexible OODBMS-IRS-coupling for structured document handling. In *Proc. 12th Intl. Conf. on Data Engineering*, S. 10–19, New Orleans, 1996.

EMMA - Eine Entwicklungsplattform für Multimediaanwendungen

Martina Schollmeyer, Günther Müller-Luschnat

Abstract

Das Projekt EMMA trägt als Ziel die Unterstützung des Entwicklers bei Entwurf und Realisierung von interaktiven World Wide Web-Anwendungen mit Multimedia-Elementen. Schwerpunkte sind dabei die Aktivitäten des *Web-Engineering* und die Realisierung von Web-Anwendungen. Geeignete Werkzeuge für diese Schritte werden in einer integrierten Plattform angeboten. Das Kernstück der Entwicklungsumgebung der EMMA (in Bezug auf das Software-Management) bildet dabei ein Repository, in dem alle im Web-Entwicklungsprozeß anfallenden Informationen und Dokumente abgespeichert sind.

1 Einleitung

Das Internet (insbesondere das World Wide Web, kurz WWW) hat in den letzten Jahren eine rasante Entwicklung vollzogen. Nachdem in den ersten Jahren der Fokus des WWW fast ausschließlich die rein passive Darstellung von Informationen war, ist in der letzten Zeit, vor allem durch die Verfügbarkeit der Programmiersprache *Java*, zu beobachten, daß WWW-Anwendungen immer stärker interaktiven Charakter bekommen. Das WWW kann dadurch zur Abwicklung geschäftlicher Transaktionen genutzt werden.

Viele Entwickler im Bereich WWW verwenden bisher keine strukturierten Methoden beim Entwurf und der Realisierung ihrer Anwendungen. Dies ist der Fall, obwohl die Methoden des Software-Engineering in Bezug auf Webanwendungen (das sogenannte Web-Engineering) bereits in der Literatur vorgestellt wurden (siehe Abschnitt 2). Das Problem ist in diesem Falle, daß das WWW ein Medium ist, in dem man schnell einfache Lösungen ohne vorhergehende Analyse produzieren und testen kann, und wo komplexe Anwendungen aus einfachen herauswachsen, die einfach weiter ergänzt werden. Nun jedoch wächst der Bedarf an Unterstützung des Webentwurfs durch strukturierte Methoden des Web-Engineering.

Der Arbeitsplatz eines WWW-Entwicklers ist durch eine Vielzahl von Werkzeugen, die verwendet werden müssen, um eine komplette Anwendung zu erstellen, gekennzeichnet. Dazu gehören HTML-Editoren, Programmierumgebungen und Bearbeitungswerkzeuge für Multimediaelemente (Video, Photo, Graphik, Text, Audio, Virtual Reality-Elemente, Animationen, etc.) [Lu94]. Um den strukturierten Entwurf von Webanwendungen zu unterstützen, müssen auch Analysewerkzeuge und Werkzeuge zur Definition des Ablaufs der interaktiven Anwendung vorhanden sein.

Das Ziel der EMMA ist die Unterstützung strukturierter Methoden bei der Entwicklung interaktiver Webanwendungen mit Multimediaelementen durch eine Integration aller benötigten Werkzeuge, vom Entwurf bis zur Realisierung, in eine einzige Plattform. Die Informationen, die während der Entwicklung einer interaktiven Anwendung entstehen, sind zum einen die eigentlichen Webseiten mit den Multimediaelementen, die dann auf einem Webserver publiziert werden. Zusätzlich entstehen Dokumente zur Informations- und Kontentanalyse, zur Informationsstrukturierung und zum Ablaufentwurf der interaktiven Anwendung. Darüber hinaus gibt es Beziehungen zwischen den Analysedaten und der letztendlich realisierten Anwendung mit ihren Komponenten, die, als Teil der Historie der Entwicklung, gespeichert werden sollten.

Für den Benutzer einer Webanwendung ist die Entwicklungsinformation im allgemeinen nicht relevant. Für den Entwickler ist es dagegen sinnvoll, eine Historie dieser Daten zu behalten. Um zwischen diesen beiden Kontexten zu trennen, wird die EMMA in zwei Umgebungen aufgeteilt: die Entwicklungsumgebung und die Produktionsumgebung.

In der Entwicklungsumgebung wird die komplette Entwicklungsinformation in einer Datenbank gehalten. Dieser separate Datenspeicher wird im Projekt EMMA durch ein Repository (Enabler von Softlab [So96]) bereitgestellt, das die Möglichkeiten der Versionsführung, des Konfigurationsmanagement und die Speicherung der gesamten Entwicklungsinformation in einem Entity-Relationship-Modell ermöglicht. Dies dient vor allen Dingen zur Unterstützung des Software-Managements. Die verschiedenen Entwicklungstools sind dabei Clients für das Repository und werden von einem gemeinsamen EMMA-Prozeß aus aufgerufen.

In der Produktionsumgebung wird die Information, die für den Benutzer der Anwendung relevant ist, d.h. die Webseiten mit den integrierten Multimediaelementen, Programmen, Hyperlinks, etc., verwaltet. Diese Information wird dann auf einem Webserver zugänglich gemacht. Ein konventioneller Webserver greift dabei nur auf Dokumente zu, die im Dokumentenbaum des Rechners gespeichert sind. Um größtmöglichen Komfort für den Benutzer bereitzustellen, wie z.B. automatische Linkverwaltung und integrierte Suchmöglichkeiten für alle auf dem Webserver gespeicherten Dokumente, können Multimediadatenbanken mit integrierter Weboberfläche als Webserver verwendet werden. Im Projekt EMMA wird als Webserver deshalb Hyperwave benutzt [Ma96]. Auf die Produktionsdatenbank wird dann mit einem Benutzer-Client (z.B. Netscape oder ein beliebiger anderer Web-Client) über HTTP (Hypertext Transfer-Protokoll) zugegriffen.

Das Ziel der EMMA ist es, die Vielzahl der hier aufgelisteten Werkzeugen miteinander zu verbinden. Die einzelnen Werkzeuge sollen dabei nicht isoliert voneinander betrachtet werden, sondern als eine Gesamteinheit. Durch das Speichern und Verwalten der Gesamtinformation in einem Repository, wird es zusätzlich möglich, solche Fragen zu beantworten, wie z.B. "Welche Webseiten sind durch Links miteinander verbunden?" oder "Welche Analyseklasse wird auf welcher Seite realisiert?" Dadurch stellt die EMMA eine integrierte Entwicklungsumgebung für interaktive Webanwendungen mit Multimediaelementen bereit.

In Abbildung 1: "Architektur der EMMA" wird die Architektur der EMMA vorgestellt. Auf der linken Seite ist die Entwicklungsumgebung mit den verschiedenen Clients dargestellt, die alle auf das Repository zugreifen. Die Aufgaben der verschiedenen Clients werden in Abschnitt 3 im Detail beschrieben. Auf der rechten Seite dagegen ist die Produktionsumgebung, in der nur der Benutzer mit einem

WWW-Client über das Internet auf die Webanwendung zugreift. Um Konsistenzprobleme zu vermeiden, ist der Informationstransfer nur von der Entwicklungs- auf die Produktionsumgebung möglich. In der Entwicklungsumgebung ist daher stets die aktuellsten Informationen gespeichert, während in der Produktionsumgebung nur die zuletzt freigegebene Version nach außen sichtbar ist, auf der auch keine Änderungen durchgeführt werden dürfen.

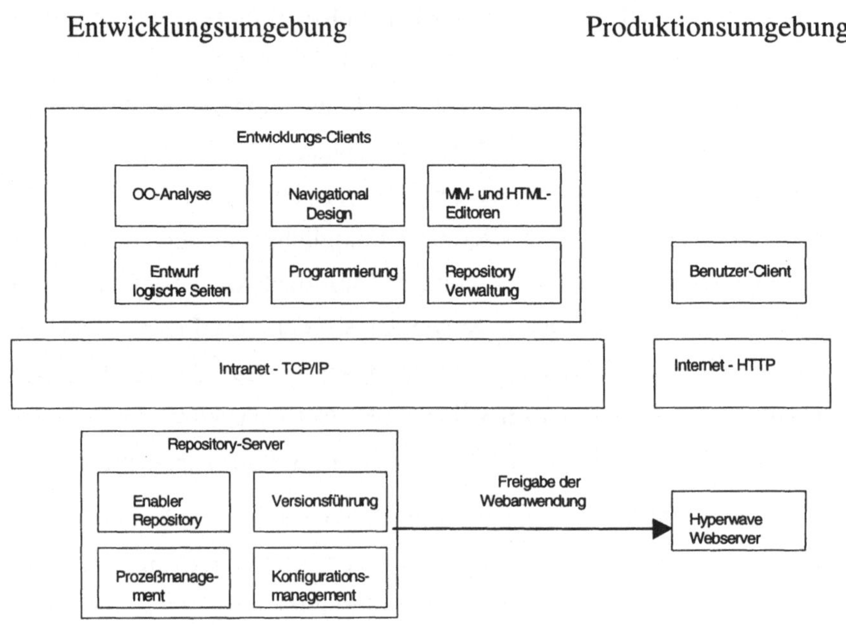

Abbildung 1: "Architektur der EMMA"

2 Die Ziele der EMMA

Auf dem Markt gibt es mehrere Werkzeuge, in denen die Erzeugung von Multimediaanwendungen bereits vollständig integriert ist (z.B. Macromedia Director, Icon Author, Multimedia Toolbook, etc.). Die meisten dieser Werkzeuge sind sogenannte Authoring-Werkzeuge, d.h., sie sind Werkzeuge, die vorprogrammierte Elemente für die Entwicklung von interaktiven Multimediaanwendungen enthalten

[MMAS]. Der Nachteil dieser Werkzeuge ist, daß Analyse und Design der Anwendung wenig unterstützt werden, daß viele der Werkzeuge als Zielplattform CD-ROM-Anwendungen haben (und kein WWW), und daß die eigene Programmierung von interaktiven Elementen meist nur wenig unterstützt wird. In der EMMA sollen aber gerade diese Ziele verfolgt werden.

2.1 Zielanwendungen

Mit der EMMA sollen interaktive WWW-Anwendungen mit Multimediaelementen erstellt werden. Der Entwickler verwendet dafür die Prozesse des Web-Engineering.

- *WWW-Anwendungen* sind gekennzeichnet durch die Verwendung des HTTP-Protokolls und der Seitenbeschreibungssprache HTML, verknüpft mit zusätzlichen Elementen.
- *Interaktive* Anwendungen erlauben dem Benutzer nicht nur Informationen aus den Webseiten abzurufen, sondern das WWW zur Durchführung von (geschäftlichen) Transaktionen zu verwenden. Dabei werden Programme (z.B. in Java) in die Webanwendung eingebunden.
- *Multimedia* bedeutet, von der Seite des Benutzers, die Verwendung von Audio und/oder Video zusätzlich zu den Standarddarstellungsformaten wie Graphiken, Photos, Text und Animationen, sowie Ergebnisse der Virtual Reality Modellierung, z.B. erstellt mit VRML [SN95].

Hypertext und Hypermedia, die die Grundlagen des WWW bilden, sind laut [SN95] ebenfalls in die Kategorie Multimedia einzuordnen. (Für eine detaillierte Beschreibung und Charakterisierung von Hypermedia siehe [Ni90].)

Die Ziele der EMMA liegen also in zwei Bereichen:

- Die Bereitstellung eines Werkzeugs zur Unterstützung der Prozesse des Web-Engineering

- Die Integration von Werkzeugen in eine einheitliche Oberfläche zur Erzeugung dieses Entwicklungs-Tools

2.2 Aktivitäten des Web-Engineering

Web-Engineering ist der Sammelbegriff für die Prozesse, die vom Entwurf bis zur Realisierung einer Webanwendung, entsprechend den Regeln des Software-Engineering, durchgeführt werden sollten. Wir befassen uns dabei mit Multimediaanwendungen im allgemeinen und mit interaktiven Webanwendungen mit Multimediaelementen im besonderen.

Laut [Ro89] besitzen Multimediaanwendungen folgende Eigenschaften:

- Sie sind auf den Benutzer fokussiert,
- sie sind Ereignis-gesteuert,
- sie werden auf einer verteilten Client/Serverarchitektur realisiert
- und mit hohen Ansprüchen in Bezug auf Performance belegt.

Die ausgewählte Methode zur Modellierung von Multimediaanwendungen sollte daher diese Bereiche abdecken können. Das heißt, es sollen Objekte, Zustände und Ereignisse modelliert werden können. Darüber hinaus sollten auch die Abläufe und ihre Steuerung durch den Benutzer und die Erfüllung der Anforderungen des Entwurfs über das Modell unterstützt werden. Diese Modellierung ist z.B. mit Jacobson's Objectory Methode [JCJÖ92] oder Rumbaugh's Objekt-Modellierungs-Technik (OMT) [RBP91] möglich. Um die Analyse von Multimediaanwendungen zu unterstützen, wird daher ein objektorientiertes Analysewerkzeug in die EMMA mit eingebunden.

Da das EMMA-Projekt sich vor allen Dingen mit der Entwicklung von Multimediaapplikationen für das WWW befaßt, sind neben den traditionellen Software-Engineering- und Multimediabegriffen auch die Begriffe und Modelle aus dem Bereich Hypermedia relevant. In der Literatur werden verschiedene Modelle verwendet, um die Entwicklung von Hypermediaanwendungen darzustellen. Beispiele dazu sind (unter anderem):

- HDM (Hypermedia Design Model) von [GPS93]
- HB1 (Hyperbase) von [SLHS93]
- RMM (Relationship Management Methodology) von [ISB95]
- OOHDM (Object Oriented Hypermedia Design Model) von [SR95],[SRB96]
- EORM (Enhanced Object Relationship Model) von [La96]

Wir konzentrieren uns hier nur auf zwei Modelle, die, unter Verwendung der Entity-Relationship- und der objektorientierten Prinzipien, die Entwicklung von interaktiven Webapplikationen unterstützen.

2.2.1 Entwicklungsschritte in RMM

Die Relationship Management Methodology basiert auf dem Entity-Relationship-Modell. Laut [ISB95] und [BMY95] werden dabei die folgenden Schritte bei der Entwicklung von Hypermedia-Anwendungen durchgeführt:

1. **Entity-Relationship-Design**: Hier werden die verschiedenen Entitäten und die Beziehungen zwischen ihnen definiert.
2. **Entity-Design**: Hier werden "Slices", also Teile der vorgegebenen Entitäten und ihrer Attribute, zusammengefaßt, die zusammen dargestellt werden sollen (z.B. auf einer Webseite). Dies ergibt die ‚logischen Webseiten‘ der Anwendung.
3. **Navigation-Design**: In diesem Schritt wird ein Diagramm erzeugt, das die verschiedenen Startpunkte, Indizes, "Rundgänge" und andere Zugangsstrukturen zur bereitgestellten Information darstellt.
4. **Conversion-Protocol-Design**: In dieser Phase wird untersucht, wie die in den früheren Schritten dargestellte Information auf einem Webserver dargestellt werden kann. Hier werden auch die späteren Werkzeuge mit in Betracht gezogen (z.B. Datenbanken, Browser, HTML-Formate, etc.)
5. **User-Interface-Design**: In diesem Schritt werden die Seiten selbst entworfen. Pro Slice (siehe Schritt 2) soll eine Seite erzeugt werden.

6. **Runtime-Behavior-Design**: Hier werden die Programme, die ausgeführt werden sollen, mit in die Seiten eingebunden. Ebenso werden hier die Links und Anker eingefügt, die auf jeder Seite aktiv sein sollen.

7. **Construction and Testing**: In diesem letzten Schritt werden die HTML-Seiten erzeugt und getestet. Hier werden auch die Zielwerkzeuge genutzt (z.B. HTML, cgi-Skripte in Perl, Embedded-SQL-Aufrufe zu Datenbanken, etc.).

2.2.2 Entwicklungsschritte im OOHDM

Bei [SRB96], [SR95], die das Objekt-Orientierte Hypermedia Design Modell (OOHDM) vorstellen, wird die Anzahl der Schritte des Web-Engineering auf vier Schritte zusammengefaßt. Dabei wird zwischen den Schritten keine feste Reihenfolge vorgeschrieben, sondern es ist jederzeit möglich, zwischen den individuellen Schritten, wie auf den Stufen einer Leiter, nach oben oder unten zu wandern.

1. **Domain-Analyse**: Hier wird die Semantik der Anwendungsdomäne modelliert. Klassen, Subsysteme, Beziehungen und Attribute werden definiert.

2. **Navigational Design**: In diesem Schritt werden die Knoten, Links, Zugriffsstrukturen, Navigationszusammenhänge, etc., definiert. Kognitive Aspekte sollen dabei berücksichtigt werden.

3. **Abstract Interface Design**: Hier werden die Objekte und die Navigationszusammenhänge miteinander verknüpft. Es werden ebenfalls die Reaktionen auf externe Ereignisse (Klicks, etc.) mit einbezogen.

4. **Implementierung**: In diesem Schritt werden die einzelnen Komponenten realisiert und die gesamte Anwendung auf ihre Vollständigkeit überprüft.

2.2.3 Web-Engineering mit der EMMA

In diesem Abschnitt beschreiben wir die Schritte, die in der EMMA unterstützt und realisiert werden sollen. Das zugrundeliegende Basismodell ist dabei das OOHDM, wobei ein Schritt des RMM eingefügt wird. Dieser Zwischenschritt, zur Strukturierung der darzustellenden Information und zur Erzeugung der logischen Webseiten, ist unserer Meinung nach ein sehr wichtiger Schritt gemäß den Aspekten des Software-Engineering, da er dem Anwender die Möglichkeit gibt, die darzustellende Information vor der Realisierung zu strukturieren und mit dem Navigational Design zu verknüpfen. Damit erhält der Entwickler bereits einen Grobüberblick über die spätere Anwendung und ihre Abläufe. Die Schritte in der EMMA sind daher:

1. **Domain-Analyse**: Die objektorientierte Modellierung wird durchgeführt
2. **Entity-Design**: Die Erstellung der logischen Webseiten als Grundlage/Templates für die spätere Realisierung
3. **Navigational Design**: Definition der Abläufe in Bezug auf die logischen Webseiten
4. **Abstract Interface Design**: Analyse und Entwurf der einzubindenden Programme und Verknüpfung mit den entsprechenden logischen Webseiten
5. **Implementierung**: Realisierung aller Entitäten der Webanwendung

2.3 Integration von verschiedenen Werkzeugen in die EMMA

In diesem Abschnitt werden die Aspekte der Integration der verschiedenen Werkzeuge in die EMMA vorgestellt und diskutiert. Die drei Dimensionen der Integration werden in der Literatur [SvdB93] als Präsentations-, Daten- und Kontrollintegration bezeichnet (presentation, information, control). Die Definition dieser drei Hauptaspekte ist wie folgt [Sc89]:

- Der Aspekt der Kontrollintegration eines Werkzeuges bestimmt seine Kommunikationseigenschaften, d.h. den Grad, zu dem es seine Ergebnisse und Aktionen anderen Werkzeugen mitteilen kann, und den Grad, zu dem andere Werkzeuge mit ihm kommunizieren können.
- Der Aspekt der Datenintegration eines Werkzeugs bestimmt den Grad, zu dem Daten, die von einem Werkzeug erzeugt werden, anderen Werkzeugen zugänglich und verständlich gemacht werden.

- Der Aspekt der Präsentationsintegration ist das Niveau, auf dem verschiedene Werkzeuge eine einheitliche Benutzeroberfläche präsentieren und sich in ähnlichen Situationen ähnlich verhalten.

Fast alle integrierten Umgebungen können mit der Kombination dieser drei relativ unabhängigen Integrationsaspekte bewertet werden.

Der Begriff Multimedia bedeutet im eigentlichen Sinne die Integration vieler Medien in eine gemeinsame Oberfläche. Das bedeutet, daß der Fokus dieser Integration im Bereich Präsentation, also bei der Benutzeroberfläche, liegt, so daß der Benutzer die Vielfalt der verwendeten Medien als eine Einheit empfindet.

2.3.1 Kategorien von Werkzeugen

Laut [SvdB93] gibt es verschiedene Kategorien von Werkzeugen, die sich darin unterscheiden, wie stark sie gekoppelt und wie kohäsiv sie sind [YC79]. Kohäsion ist dabei ein Maß für die Stärke der Argumente, warum bestimmte Entitäten zusammen gruppiert werden sollen. Kopplung dagegen befaßt sich damit, wie stark die Entitäten, die man nicht zusammengruppiert, voneinander abhängig sind. In den im folgenden aufgelisteten Kategorien nimmt dabei von Schritt zu Schritt das Maß der Kohäsion zu und das der Kopplung ab.

1. Ein *Werkzeug* ist eine Sammlung von Diensten, die intern sehr kohäsiv sind und die nach außen wenig gekoppelt sind. Beispiele für Werkzeuge sind z.B. Editoren, Compiler, etc.
2. Eine "*Werkzeugkiste*" (Toolset) ist eine Sammlung von Werkzeugen, die intern sehr kohäsiv sind und die nach außen wenig gekoppelt sind. Ein Beispiel hierfür ist ein Compiler mit dem dazugehörigen Debugger - zwei getrennte Werkzeuge, die trotzdem eng zusammenarbeiten.
3. Eine *Umgebung* (Plattform) ist eine Sammlung von Toolsets, die intern sehr kohäsiv ist und nach außen wenig gekoppelt. Hierfür soll die EMMA ein Beispiel sein - die Integration verschiedener Werkzeuge in eine gemeinsame Oberfläche.

2.3.2 Integration in der EMMA

In der EMMA sind zwei Integrationsaspekte besonders relevant: die Präsentationsintegration und die Datenintegration. Die Kontrollintegration wird dabei vernachlässigt, da alle Informationen, die in der EMMA bearbeitet werden, von einem Repository verwaltet werden. Dieses Repository kann von verschiedenen Benutzern gleichzeitig verwendet werden und enthält deshalb bereits die nötigen Kontrollaspekte.

2.3.2.1 Datenintegration

Die Datenintegration wird durch die Verwendung des Repository unterstützt. Damit diese Integration stattfinden kann, müssen die verschiedenen Werkzeuge, die integriert werden sollen, ein gemeinsames Verständnis für die gemeinsam verwendeten Daten haben. Dies wird normalerweise durch ein Datenmodell realisiert, in dem Objekttypen, Operationen und Konstruktionsregeln definiert werden [SvdB93]. Für das Datenmodell werden häufig Entity-Relationship- oder objektorientierte Modelle verwendet. Für das EMMA-Projekt wird in einem gesonderten Bericht [SMW97] das für die Datenintegration notwendige erzeugte Daten- bzw. Metamodell vorgestellt.

Neben dem Datenmodell sind auch noch andere Aspekte wichtig für die Datenintegration. Hierzu gehören, unter anderem, auch Versions- und Konfigurationsmanagement [BEM91]. Das verwendete Repository [So96] unterstützt bereits all diese Aspekte, so daß von seiten der EMMA die Datenintegration hauptsächlich durch das im Repository festgelegte Datenmodell geschieht.

2.3.2.2 Präsentationsintegration

Um die Entwicklungsplattform der EMMA zu erzeugen, müssen verschiedene Werkzeuge miteinander integriert werden. Die Grundlage der Entwicklungsplattform jedoch fordert, daß die Benutzeroberfläche für die eingebundenen Werkzeuge so einheitlich wie möglich gestaltet wird. Die Benutzeroberfläche sollte dabei jedoch von der eigentlichen Anwendung getrennt betrachtet werden.

Bei der Entwicklung des EMMA-Prototyps werden Werkzeuge (sowohl bereits existierende als auch selbst entwickelte) für die einzelnen Entwurfs- und Realisierungsschritte in der EMMA integriert. Da für die bereits existierenden Werkzeuge die Benutzeroberfläche bereits gestaltet wurde und normalerweise nicht mehr verändert werden kann, können in der EMMA in Bezug auf die Präsentation nur noch die Werkzeuge beeinflußt werden, die im Rahmen des Projekts selbst entwickelt werden.

Die Benutzeroberfläche des EMMA-Clients dient also fast ausschließlich nur als Oberfläche zum Aufruf der anderen, integrierten, Werkzeuge. Diese Werkzeuge im EMMA-Client sind, von der Benutzung her, bereits relativ einheitlich aufgebaut, da sie alle auf der gleichen Plattform laufen (in unserem Fall Windows '95). Anstatt jedoch, wie gewohnt, ihre Dateien auf dem Client-Rechner selbst zu verwalten, werden die Dateien von den jeweiligen Programmen, relativ transparent für den Benutzer, im Repository abgelegt und verwaltet. Die zusätzlich benötigten Funktionen werden in Visual Basic programmiert, so daß zumindest eine einheitliche Gestaltung der Oberfläche dieser Funktionen erreicht werden kann.

3 Die Entwicklungsumgebung der EMMA

In diesem Abschnitt wird die Entwicklungsumgebung der EMMA vorgestellt. Zum einen wird das Systemumfeld festgelegt und zum anderen werden die einzelnen zu integrierenden Komponenten vorgestellt.

3.1 Systemumfeld

Die EMMA wird als Client-Server-Anwendung realisiert. Als Server dienen dabei das Repository und der Hyperwave-Webserver. Wie bereits in Abbildung 1 gezeigt wurde, sind die Entwicklungsumgebung, mit dem Repository als Server, und die Produktionsumgebung, mit dem Webserver als Server, fast völlig getrennt realisiert. Die verschiedenen Anwendungsprogramme, die im Rahmen der Web-Anwendungserstellung auf das Repository zugreifen, werden in einem EMMA-Client integriert. Der Client für den Webserver ist nicht vorgeschrieben.

- Als Client-Betriebssystem für den EMMA-Client wird Microsoft Windows '95 verwendet.

- Als Server-Betriebssystem wurde UNIX gewählt.

Die für die EMMA notwendige Programmierung wird mit Visual Basic durchgeführt, da die verwendeten Werkzeuge, insbesondere das Repository und das ausgewählte Analyse-Werkzeug Visual Basic-Schnittstellen aufweisen.

3.2 Komponenten

Die EMMA wird alle Werkzeuge beinhalten, die ein Webentwickler für seine Arbeit benötigt. Ein Grundsatz ist, daß die hier einzusetzenden Werkzeuge am Markt erhältlich sind und nur diejenigen Entwicklungsprozesse realisiert werden, für die es noch keine fertigen Werkzeuge gibt. Um die spezifischen Werkzeuge zu bestimmen, die in der EMMA integriert werden sollen, wurde für jede der in Abbildung 2 aufgelisteten Komponenten gemäß von Kriterienlisten zur Integration und zu den gewünschten technischen Eigenschaften ein Auswahlprozeß durchgeführt. Die integrierten/zu integrierenden Komponenten sind (siehe Abbildung 2):

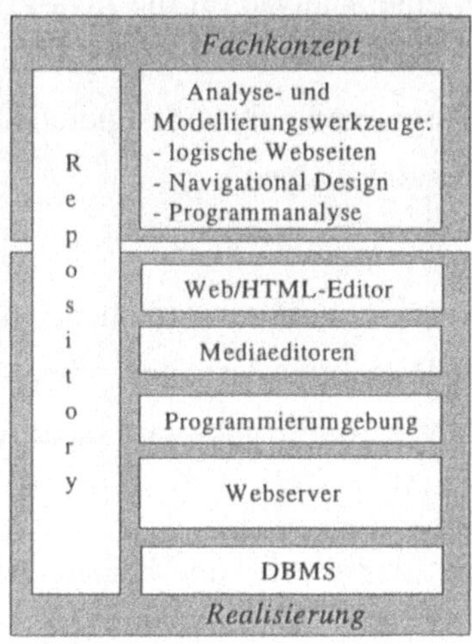

Abbildung 2 : "EMMA-Komponenten"

- *Analysewerkzeuge*

  Wie aus den Aktivitäten des Web-Engineering zu ersehen ist, wird zuerst eine (objektorientierte) Analyse durchgeführt. Viele Autoren empfehlen die Analyse auch zur Strukturierung der Informationsinhalte, die auf den Webseiten präsentiert werden. Eine weitere wichtige Methode in der Analyse ist die Prozeßmodellierung. Das für die EMMA ausgewählte Werkzeug SELECT Enterprise unterstützt sowohl OO-Analyse als auch Prozeßmodellierung.

- *Werkzeug zur Modellierung der logischen Webseiten*
  Dieses Werkzeug wird im Rahmen der EMMA neu entwickelt. Dazu wird ein Texteditor zum Beschreiben der Inhalte der logischen Webseite mit zusätzlicher Funktionalität ausgestattet, um Verknüpfungen zwischen Objekten im Repository herzustellen. Die Programmierung erfolgt mit Visual Basic.

- *Werkzeug zum Modellierung eines Navigational Design*
  Dieses Werkzeug wird ebenfalls im Rahmen der EMMA neu entwickelt. Die Funktionalität besteht dabei aus der graphischen Verknüpfung von logischen Webseiten mit Verweisen. Die Programmierung erfolgt in Visual Basic.

- *Werkzeug zur Programmanalyse*
  Mit diesem Werkzeug soll eine Analyse für die Java-Programmierung durchgeführt werden. Da die Java objektorientiert ist, ist auch die Analyse objektorientiert. Damit kann für die Programmanalyse das gleiche Werkzeug verwendet werden wie für die Informationsanalyse.

- *Web/HTML-Editor*
  Um die Erstellung der Webseiten zu vereinfachen, wird ein Tool benötigt, das gemäß dem WYSIWYG-Paradigma arbeitet, aber auch das Editieren von HTML-Sourcecode ermöglicht. Im Projekt wird dazu AOLPress verwendet.

- *(Multi)-Mediaeditoren*
  Mit Mediaeditoren werden hier Werkzeuge bezeichnet, mit denen (unter anderem) die Medien Photo, Graphik, Audio und Video bearbeitet werden können. Die Auswahl der meisten Werkzeuge ist dabei bereits getroffen: Corel Draw für Graphiken, Picture Publisher für die Photobearbeitung und Adobe Premiere zur Videobearbeitung.

- *Programmierumgebung*

  In die EMMA wird eine Programmierumgebung für die Programmiersprache Java eingebunden (Microsoft Visual J++).

- *Webserver*

  Als Produktionsplattform wird ein Webserver benötigt. In der EMMA wird hierfür das Produkt Hyperwave verwendet. Hyperwave ist ein Webserver der "nächsten Generation" [Ma96], da die herkömmliche Funktionalität eines Webservers durch eine integrierte Multimediadatenbank mit Webtechnologie ergänzt wird Durch diese integrierte Datenbank besitzt Hyperwave eine integrierte Linkverwaltung und Volltextsuche auf allen gespeicherten HTML-Dokumenten und die Möglichkeit, alle gespeicherten Dokumente mit Schlüsselwörtern und Annotationen zu versehen.

- *Repository*

  Zusammengeführt werden alle Komponenten über eine integrierte Ergebnisablage in einem Repository. Hiermit ist es möglich, Konsistenz zwischen den Ergebnistypen verschiedener Werkzeuge zu gewährleisten und Versionierung und Prozeßmanagement zu vereinfachen. Als Produkt wird hier das Repository Enabler der Firma Softlab verwendet.

4 Fazit

Die wenig genutzte Disziplin des Web-Engineering macht eine integrierte Entwicklungsplattform für interaktive Multimediaanwendungen für das WWW wie die EMMA eine sehr interessante Option für Entwickler. EMMAs großer Vorteil liegt vor allem in der integrierten Entwicklungsumgebung basierend auf einem Repository, das Versionsführung und Konfigurationsmanagement ermöglicht. Die Möglichkeit, eine Historie der Entwicklung zu behalten und diese getrennt von einer Produktionsumgebung zu halten, macht die EMMA für den Bereich Softwaremanagement besonders interessant. Dadurch können auch im nachhinein die Schritte der Entwickler nachvollzogen werden.

Literatur

[BMY95] Balasubramanian, V., Ma, B.M., & Yoo, J. A Systematic Approach to Designing a WWW Application. *Communications of the ACM*, Vol. 38, No. 8, 1995, S.47-48.

[BEM91] Belkhatir, Estbublier & Melo. Adele2: A Support to Large Software Development Process. *Proceedings of the 1st International Conference on the Software Process*, Redondo Beach, CA, USA, Oktober 1991.

[GPS93] Garzotto, F., Paolini, P. & Schwabe, D. HDM - A Model-Based Approach to Hypertext Application Design. *ACM Transactions on Information Systems*, Vol. 44, No. 1, 1993, S.1-26.

[ISB95] Isakowitz, T. Stohr, E.A. & Balasubramanian, P. RMM: A Methodology for Structured Hypermedia Design. *Communications of the ACM*, Vol. 38, No. 8, 1995, S.34- 44

[JCJÖ92] Jacobson, I., Christerson, M., Jonsson, P. & Övergaard, G.*: Object-Oriented Software Engineering - A Use Case Driven Approach.* ACM Press, 1992.

[Lu94] Luther, A.C. *Authoring Interavtice Multimedia.* Academic Press, 1994.

[La96] Lange, D. An Object-Oriented Design Approach for Developing Hypermedia Information Systems, *Journal of Organizational Computing*, Vol. 6, No. 3, 1996. S.269-293.

[Ma96] Maurer H. Hyperwave: *The next generation web solution.* Addison-Wesley, 1996.

[MMAS] Multimedia Authoring Software Frequently Asked Questions. http://www.tiac.net/users/jasiglar/MMASFAQ.HTML

[Ni90] Nielsen, J. *Hypertext and Hypermedia.* Academic Press, 1990.

[RBP91] Rumbaugh, J., Blaha, M., Premerlani, W., Eddy, F. & Lorensen, W.*: Object-Oriented Modeling and Design.* Prentice Hall 1991.

[Ro89] Rosenberg, D. Applying Object-Oriented Methods to Interactive Multimedia Projects. *Object Magazine*, Vol. 5, No. 3, Juni 1995, S. 41-47.

[Sc89] Schefström, D. Building a highly integrated environment using preexisting parts. In *IFIP'89*, San Francisco, CA, USA, September 1989. North-Holland.

[SvdB93] Schefström, D. & van den Broek, G. (Editors) *Tool Integration: Environments and Frameworks*. John Wiley and Sons, 1993.

[SLHS93] Schnase, J.L., Legget, J.J., Hicks, D.L. & Szabo, R.L. Semantic Data Modelling of Hypermedia Associations. *ACM Transactions on Information Systems*, Vol. 44, No. 1, 1993, S.27-50.

[SMW97] Schollmeyer, M., Müller-Luschnat, G. & Wilke, M. EMMA Fachkonzept. *FAST Technischer Bericht* 97-01, März 1997.

[SR95] Schwabe, D. & Rossi, G. The Object-Oriented Hypermedia Design Model. *Communications of the ACM*, Vol. 38, No. 8, 1995, S.45-46.

[SRB96] Schwabe, D., Rossi, G., & Barbosa, S.D.J. Systematic Design of Hypermedia Applications using OOHDM. *Proceedings of the ACM Conference on Hypertext, Hypertext'96*, Washington DC, März 1996.

[So96] Softlab GmbH. *Enabler Repository Reference Guide*. Dezember 1996.

[SN95] Steinmetz, R. & Nahrstedt, K. *Multimedia: Computing, Communications and Applications*. Prentice-Hall, 1995.

[YC79] Yourdon & Constantine. *Structured Design*. Prentice-Hall, 1979.

Abschlußbericht der GI-Arbeitsgruppe "Vergleichende Analyse von Problemstellungen und Lösungsansätzen in den Fachgebieten Information Systems Engineering, Software Engineering und Knowledge Engineering"[1]

Jürgen Angele, Rudi Studer

Abstract

Im folgenden wird das Ergebnis einer Arbeitsgruppe dargestellt, deren Ziel der Vergleich unterschiedlicher Problemstellungen und Lösungsansätze in den Bereichen Information Systems Engineering, Software Engineering und Knowledge Engineering war. Dazu werden beispielhaft vier unterschiedliche Methodiken aus diesen Bereichen kurz charakterisiert. Anschließend werden die Unterschiede in den Problemstellungen und Methoden dieser Ansätze, aber auch deren Gemeinsamkeiten aufgezeigt. Es zeigt sich, daß die Methodiken aus den jeweiligen Gebieten unterschiedliche Stärken aufweisen, so daß es angebracht erscheint, die unterschiedlichen Komponenten bei komplexen Problemstellungen jeweils auch mit den dafür geeignetsten Methoden aus den drei Fachgebieten zu entwickeln. Dies erfordert dann auch eine größere Kommunikations- und Kooperationsbereitschaft zwischen diesen Gebieten und eine gemeinsame Behandlung der Methoden in der Lehre.

1 Einleitung

Mit Fragen der methodischen Unterstützung des Entwicklungsprozesses von Soft-

---

[1] Dies ist eine verkürzte Fassung des ausführlichen Abschlußberichtes (siehe [APS+97])

waresystemen beschäftigen sich verschiedene Teildisziplinen der Informatik. Speziell sind hier das Software Engineering, das Information Systems Engineering und das Knowledge Engineering zu nennen. Während das Software Engineering insbesondere Beiträge zur Beschreibung des Entwicklungsprozesses durch Vorgehensmodelle und zur Beschreibung (nicht-)funktionaler Aspekte von Softwaresystemen geliefert hat, beschäftigte man sich im Information Systems Engineering zunächst primär mit der Modellierung statischer Aspekte von Informationssystemen durch semantische Datenmodelle. In den zurückliegenden Jahren gewannen dynamische Aspekte jedoch immer mehr an Bedeutung. Im Knowledge Engineering wurden ursprünglich hauptsächlich Fragen der methodischen Untersützung der Wissenserhebung untersucht, in jüngster Zeit bekamen jedoch Methoden zur Wissensmodellierung und -wiederverwendung ein immer stärkeres Gewicht.

Bei einer näheren Betrachtung dieser Fachgebiete zeigt es sich, daß einerseits eine Vielzahl gemeinsamer Fragestellungen und Lösungsansätze existieren, andererseits aber auch sehr unterschiedliche Problemstellungen untersucht werden.

Um das gegenseitige Verständnis zwischen diesen verschiedenen Fachgebieten zu verbessern, wurden von mehreren GI-Fachgruppen gemeinsam in den Jahren 1990, 1992 und 1994 drei Workshops "Informationssysteme und Künstliche Intelligenz" veranstaltet (siehe [Kar90], [Stu92], [LuM94]). Auf diesen Workshops zeigte es sich jedoch, daß ein eingehendes Verstehen der jeweils anderen Disziplinen und ein fundiertes Herausarbeiten von Gemeinsamkeiten und Unterschieden im Rahmen von einzelnen, in zweijährigem Abstand stattfindenden Workshops mit jeweils wechselnden Teilnehmern nur schwer zu erreichen ist. Deshalb beschlossen die GI-Fachgruppen Knowledge Engineering (FG 1.5.1), Software Engineering (FG 2.1.1) und Entwicklungsmethoden für Informationssysteme und deren Anwendung (EMISA) (FG 2.5.2) 1992 die Einrichtung einer gemeinsamen Arbeitsgruppe. Ziel der Arbeitsgruppe sollte es sein, "Problemstellungen und Lösungsansätze in den Fachgebieten Information Systems Engineering, Software Engineering und Knowledge Engineering vergleichend zu analysieren, um so Gemeinsamkeiten und Unterschiede besser charakterisieren zu können."

Die Arbeitsruppe konstituierte sich im Rahmen des Workshops "Querbezüge des

Knowledge Engineering zu Methoden des Software Engineering und der Entwicklung von Informationssystemen" auf der 2. Deutschen Tagung Expertensysteme [AnS93]. Anfangs beteiligten sich zehn verschiedene Gruppen bzw. Einzelpersonen an der Arbeitsgruppe (ein Überblick ist in [ALO+93] zu finden. Zur Fokussierung der Arbeiten beschloß die Arbeitsgruppe, sich primär mit den Themen Vorgehensmodelle und Methoden zu beschäftigen. Unter einem Vorgehensmodell wurde dabei die „Festlegung der bei der Entwicklung eines Systems durchzuführenden Arbeitsschritte verstanden, ... Beziehungen zwischen den Arbeitsschritten sind ebenso festzulegen wie Anforderungen an die zu erzeugenden Ergebnisse.“ [ALO+93]. Als eine Methode wurde eine "systematische Handlungsvorschrift zur Lösung von Aufgaben einer bestimmten Art verstanden." [ALO+93]. Dementsprechend wurde in der Arbeitsgruppe der Begriff Methodik im Sinne von Methodensammlung verwendet.

Außerdem einigte man sich in der Arbeitsgruppe darauf, die Arbeiten anhand einer vergleichenden Fallstudie durchzuführen. In Abwandlung des oft verwendeten IFIP Beispiels [OSV82] wurde als Aufgabenstellung für die Fallstudie die Entwicklung eines (wissensbasierten) Systems zur Tagungsverwaltung ausgewählt.

Im Rahmen ihrer Arbeit organisierte die Arbeitsgruppe noch einen weiteren Workshop "Vorgehensmodelle und Methoden zur Entwicklung komplexer Softwaresysteme", der auf der 18. Deutschen Jahrestagung für Künstliche Intelligenz durchgeführt wurde [KuS94].

Leider zeigte es sich in der laufenden Arbeit der Arbeitsgruppe, daß es insbesondere für Mitglieder aus der Wirtschaft sehr schwierig ist, sich über eine längeren Zeitraum aktiv an einer derartigen Arbeitsgruppe zu beteiligen. So blieben für die letzte Phase der Arbeitsgruppe nur noch vier Gruppen übrig, die auch in dieser Kurzfassung des Abschlußberichts vertreten sind. Von daher sollte klar sein, daß dieser Abschlußbericht keine alle Aspekte umfassende Analyse sein kann, sondern sich vielmehr auf Schlußfolgerungen beschränken muß, die auf Grund der analysierten Methodiken möglich sind. Gleichwohl beinhalten diese Methodiken aus Sicht der Autoren typische methodische Vorgehensweisen in den beteiligten Fachgebieten. Um einen systematischen Vergleich der Methodiken zu ermöglichen, erarbeitete die

Arbeitsgruppe einen Kriterienkatalog, mit dem charakteristische Eigenschaften einer Methodik erfaßt werden können [Kri97].

2 Die gewählten Vorgehensweisen in der Arbeitsgruppe

Nachfolgend werden die Vorgehensweisen sowie die Methoden und Werkzeuge, die die Mitglieder der Arbeitsgruppe für die Bearbeitung der Fallstudie einsetzten, kurz zusammengefaßt:

MIKE

Im Projekt MIKE (Modellbasiertes und Inkrementelles Knowledge Engineering) [AFS96] wird eine Methodensammlung zum Bau von Expertensystemen entwickelt und ein prototypisches Werkzeug konzipiert und realisiert. Dazu werden die Vorteile von Vorgehensmodellen, formalen Spezifikationssprachen und Prototyping in einen homogenen Ansatz integriert. Im Prozeßmodell von MIKE wird der Entwicklungsprozeß in die Phasen Akquisition, Design, Implementierung und Evaluierung unterteilt, die zyklisch durchlaufen werden, um das System inkrementell zu entwickeln. Die Akquisition wird ihrerseits in die Teilschritte Aufgabenanalyse, Modellbildung und Modellevaluierung unterteilt, die auch zyklisch durchlaufen werden, so daß das Ergebnis der Akquisition, das Modell der Expertise, inkrementell durch exploratives Prototyping erstellt werden kann. Das Modell der Expertise wird einerseits mit einer semiformalen Darstellung basierend auf Hypertext [Neu93], und mit der formalen und ausführbaren Sprache KARL [FAS97] beschrieben. Das Modell der Expertise besteht aus den drei Ebenen: Kontrollebene, Inferenzebene und Domänenebene. Auf der Kontrollebene wird der Kontrollfluß, auf der Inferenzebene der Datenfluß und auf der Domänenebene das fachspezifische Wissen beschrieben. Die Kontrollebene und Inferenzebene zusammen beschreiben das dynamische Verhalten des Systems in Form einer Problemlösungsmethode. Eine solche Problemlösungsmethode kann generisch, also unabhängig vom fachspezifischen Wissen der Domänenebene formuliert werden. Dies ermögliche deren Wiederverwendung in einem anderen Anwendungsgebiet und bildet die

Grundlage zur Entwicklung von Bibliotheken von Problemlösungsmethoden. Zur Modellierung statischen Wissens auf der Domänenebene stellt die Sprache KARL Konzepte aus objektorientierten Datenmodellen zur Verfügung. Die Spezifikation regelhafter Zusammenhänge auf der Domänenebene sowie der Ein- und Ausgabebeziehungen von Prozessen auf der Inferenzebene erfolgt in KARL durch Regeln. Diese basieren auf Hornregeln mit Negation, die um objektorientierte Primitive erweitert wurden. Der Kontrollfluß wird durch prozedurale Konstrukte wie Sequenz, Fallunterscheidung und Wiederholung spezifiziert. Ein in KARL beschriebenes Modell der Expertise ist direkt ausführbar und läßt sich somit durch Prototyping entwickeln. Ausgehend von der Beschreibung der funktionalen Anforderungen im Modell der Expertise konzentriert man sich dann in einer Design-Phase auf Realisierungsaspekte und nicht-funktionale Anforderungen wie Effizienz, Wartbarkeit usw. Die zu entwickelnden unterschiedlichen Modelle (informales Modell der Expertise, formales Modell der Expertise, Designmodell, wissensbasiertes System) erlauben eine schrittweise Überführung von informalem Wissen in die endgültige Implementierung des Systems. Zur besseren Nachvollziehbarkeit werden Querbezüge zwischen den Modellen explizit repräsentiert. Das MIKE-Tool enthält eine Reihe graphischer Editoren zur Entwicklung der Modelle und einen Interpreter/Debugger für KARL.

INCOME/STAR

Ziel des INCOME/STAR-Projekts [OSS94] war die Konzeption und prototypmäßige Implementierung einer kooperativen Entwicklungs- und Wartungsumgebung für verteilte Informationssysteme. Das Vorgehensmodell ProMISE (Project Model for Information System Evolution) unterscheidet die Phasen Planung, Anforderungsdefinition und -analyse, Entwurf, Implementation, Test und Dokumentation, Installation, Einsatz und Wartung. Diese Phasen werden zyklisch bearbeitet. Jede Phase enthält u.a. einen phasenbezogenen Planungsschritt sowie Validierungs- und Verifizierungsschritte. Diese Schritte sind selbst ebenfalls zyklisch zu durchlaufen. Jede Phase beginnt mit einem - so weit wie möglich formalen - Transformationsschritt, bei dem die Resultate der vorhergehenden Phasen in (initiale) Resultate der aktuellen Phase überführt werden. In der Phase der Infomationsbedarfsermittlung und -analyse wird eine semiformale Darstellung der statischen Strukturen und dynamischen Abläufe erstellt. Die Informationen werden nach Daten-, Operations- und Ereignisanforderungen klassifziert und in Glossaren abgelegt. Zur anschließenden Datenmodellierung wird entweder das semantisch-hierarchische Objektmodell SHM oder eine Variante des Entity/Relationship-Modells benutzt. Ein kontrollierter Übergang zwischen beiden Modellen wird methodisch unterstützt. Zur Reduktion der Komplexität werden beim Entity/Relationship-Modell Clustering-Techniken verwendet. Dabei werden größere Ausschnitte aus einem detaillierten Schema auf sog. Entity-Cluster abgebildet. Das Systemverhalten wird mit Hilfe eines neuartigen Petri-Netz-Typs, den sog. NR/T-Netzen, definiert. Den Stellen im Netz werden Datenschemata zugeordnet. NR/T-Netze erlauben verteilte Abläufe und komplexe Objektstrukturen in integrierter Form zu modellieren. Ein in dieser Form beschriebenes Modell ist direkt ausführbar und kann somit durch Prototyping entwickelt werden. Als Werkzeuge werden INCOME (Fa. PROMATIS [INC93]) bzw. INCOME/STAR (Weiterentwicklung Uni Karlsruhe [NOS92]) und ORACLE*CASE eingesetzt. Diese stellen Editoren für die verschiedenen Diagramm-Typen, Generatoren (z.B. DB-Generator, Source-Code-Generator), Simulator für Pr/T-Netze sowie Analysewerkzeuge bereit.

CoMo-Kit

Im Projekt CoMo-Kit [DMP97] werden Methoden, Techniken und Werkzeuge für ein Workflowmanagementsystem entwickelt. Dieses erlaubt die Modellierung von flexiblen und wissensintensiven Arbeitsabläufen und unterstützt deren Abwicklung. Ausgehend von natürlichsprachlichen Protokollen werden Prozeß- und Produktmodelle entwickelt. Prozessmodelle sind durch Prozesse, Methoden und Parameter charakterisiert. Produktmodelle enthalten Schemabeschreibungen für Daten und konkrete Instanzen dieser Schemabeschreibungen. Eine Workflowmaschine erlaubt die Simulation bzw. Ausführung eines Projektes, so daß hierdurch die Entwicklung der Modelle durch Prototyping unterstützt wird. Die Workflowmaschine erhält die Prozeßmodelle als Eingabe und unterstützt und koordiniert die Abwicklung des zugehörigen Projektes. Dazu verwaltet diese den aktuellen Projektzustand, die im Projektlauf entstandenen Produkte, Arbeitslisten für die Benutzer und die Weiterleitung von Informationen an den Benutzer. Jede im Prozeßmodell enthaltene Aufgabe wird durch Zuordnung von WWW-Seiten beschrieben, die im Verlauf der Systementwicklung in Zusammenarbeit mit Anwendungsexperten erstellt wurden. Im Projekt CoMo-Kit wurden Sprachen zur Beschreibung des Prozeß- und des Produktmodells entwickelt. Prozesse werden darin durch ihre Ziele, Parameter, das verfügbare Hintergrundwissen, Vor- und Nachbedingungen, durch die Anforderungen an Aufgabenbearbeiter und durch eine Liste von Methoden beschrieben. Methoden beschreiben dann, wie das Ziel eines Prozesses erreicht werden kann. Methoden werden entweder durch menschliche Agenten oder durch Computeragenten ausgeführt. Methoden sind entweder atomar oder zerlegen einen Prozeß in einfacher zu lösende Teilprozesse. Produktdaten werden in CoMo-Kit objektorientiert modelliert. Produkttypen sind beschreibbar duch eine Menge von Slots. Agenten werden durch deren Qualifikationen, Rollen und der organisatorischen Einordnung beschrieben. Das entwickelte Werkzeug enthält graphische Editoren zur Entwicklung der Modelle und die Realisierung der Workflowmaschine.

MORE/RT

Die objektorientierte Analysemethode MORE/RT [Ste95] ist insbesondere zur Anforderungsdefinition von verteilten, nebenläufigen Echtzeitsystemen geeignet. Sie hebt sich von anderen objektorientierten Analysemethoden durch eine bessere Beschreibbarkeit der Kommunikation und Synchronisation der Systemkomponenten sowie von zeitlichen Anforderungen ab. Es werden die Phasen Istanalyse, Anforderungsdefinition, Entwurf und Implementierung unterschieden. Die ersten beiden Phasen werden direkt durch MORE/RT, die letzten beiden durch die Methode COOD [Hüs95] unterstützt. In einem MORE/RT-Systemmodell beschreiben Klassendiagramme, Vererbungsstrukturen, Aggregationsstrukturen, Assoziationsstrukturen, Kommunikationsstrukturen und Inkarnationsstrukturen. Methoden von Klassen werden entweder natürlichsprachlich oder, falls das Modell simuliert werden soll, in der Sprache MORE/RT-PC formuliert. Das dynamische Verhalten eines

Modells wird durch sog. Objektzustandsgraphen beschrieben. Sie basieren auf Statecharts und deren Weiterentwicklung zu Objectcharts. Objektdiagramme beschreiben die statischen Beziehungen zwischen typischen Instanzen, wie sie während des Systembetriebs auftreten können. Für Methoden kann deren Verarbeitungszeit und für Botschaften deren Laufzeit spezifiziert werden. Diese Zeitabschätzungen bilden die Grundlage zu einer worst-case-Kalkulation der Antwortzeiten des Gesamtsystems. Diese kann durch die Simulation des Systemmodells überprüft werden. Das Zeitverhalten wird durch sog. Zeitdiagramme, dargestellt. Ein Zeitdiagramm ist ein zweidimensionales Diagramm, bei dem auf einer Achse die Simulationszeit und auf der anderen Achse zu betrachtende Objekte aufgetragen werden. Die Simulation eines Modells wird auf ein sog. Szenario angewandt. Ein Szenario beschreibt einen typischen Anwendungsfall. Es ist durch die Anfangszustandsbelegung der einzelnen Objekte, die Liste der externen Botschaften und evtl. durch eine Terminationsbedingung charakterisiert.

3 Vergleich der Vorgehensweisen

Die Lehrbücher [Som92], [SoK93], [DKS93] bzw. die einschlägige Literatur der drei Gebiete Software Engineering, Information Systems Engineering und Knowledge Engineering drücken im allgemeinen eine strikte Trennung der drei Gebiete aus. So findet man darin nur sehr selten Querverweise zu den jeweils benachbarten Gebieten. Dies hat zum einen historische Ursachen und zum anderen liegt dies sicher in der oft sehr strikten Trennung der Gruppen, die jeweils in den unterschiedlichen Gebieten Methoden entwickeln und einsetzen, begründet. So existieren weder auf der Forschungsseite, noch auf der Entwicklungsseite, noch auf der Anwenderseite etablierte gemeinsame Foren, in denen Erkenntnisse der unterschiedlichen Gruppen ausgetauscht werden könnten.

In die betriebliche Praxis finden überwiegend Methoden des Software Engineering Eingang. In speziellen Fällen werden auch Methoden des Information Systems Engineering, z.B. zur Beschreibung betrieblicher Abläufe oder zum Entwurf von betrieblichen Informationssystemen, angewandt. Methoden des Knowledge Engineering haben, auch zum Bau wissensbasierter Systeme, bisher praktisch keinen Eingang in die betriebliche Praxis gefunden. Hier werden Systeme also noch am öftesten weitgehend auf der Codeebene, also ohne Einsatz einer Methodik für die Analyse und den Entwurf, entwickelt. Allerdings ist in der Praxis auch in den anderen Gebieten noch keine vollständige Durchdringung des Einsatzes von entsprechenden Methoden zu verzeichnen.

Natürlich weisen die einzelnen Methodiken, und das zeigte sich auch im Verlaufe der Arbeit dieser Arbeitsgruppe, Charakteristiken auf, die speziell auf die jeweiligen Anwendungsgebiete zugeschnitten sind.

So erlauben Methoden des Software Engineering im Gegensatz zu den Methoden der anderen Gebiete die Entwicklung und Beschreibung komplexer Algorithmen. Oft lassen sich Realzeitaspekte, wie sie z.B. für Systeme zur Prozeßkontrolle benötigt werden, beschreiben. Bei Methodiken, die eng mit der Strukturierten Analyse verwandt sind, wird ein System ausgehend von den Prozessen und Algorithmen, also der Dynamik des Systems, entwickelt. Bei objektorientierten Methodiken stehen dagegen die Daten und deren Struktur im Vordergrund. Letztere bieten deshalb auch umfangreichere Methoden zur Beschreibung der statischen Strukturen. Diese sind deshalb näher verwandt mit Methoden des Information System Engineering und gingen sogar zum Teil aus deren Methoden hervor. Im Bereich der statischen Modellierung ist deshalb eine Trennung zwischen den Methoden des Information System Engineering und objektorientierten Methoden weitgehend aufgehoben. Objektorientierte Methoden eignen sich sehr gut zur Entwicklung diskreter Systeme, bei denen in jedem Schritt die Kenntnis der Zustände nur sehr weniger Objekte notwendig ist und bei denen die Ergebnisse nur auf einfache Art mit den auftretenden Ereignissen verknüpft sind. Strukturierte Ansätze scheinen dagegen besser für komplexe Algorithmen, für Probleme, bei denen das Resultat vom Zusammenspiel vieler Objekte (z.B. globale Optimierungsprobleme) abhängt, und für Probleme hoher kombinatorischer Komplexität (z.B. Zuordnungsprobleme) geeignet zu sein. In den gängigen Software Engineering Methoden werden bis auf das realisierte System nur semiformale Sprachen zur Beschreibung von Modellen eingesetzt. Dies hat zur Konsequenz, daß solche Modelle weder formalen Methoden, wie z.B. Verifikationsmethoden, zugänglich sind, noch daß sie zu ihrer Validierung ausgeführt werden können.

Methoden des Information System Engineering bieten insbesondere ausgefeilte Methoden zur Beschreibung der statischen Struktur (Schemabeschreibungen) der Daten an. Modellierungsprimitive aus den semantischen Datenmodellen ermöglichen die Beschreibung vielfältiger Beziehungstypen zwischen einzelnen Klassen (Entitytypen) von Objekten. Bei der Beschreibung der Dynamik von Systemen stehen hier oft Verteilungsaspekte im Vordergrund, wie sie z.B. zur Modellierung betrieblicher Abläufe benötigt werden. Damit einher gehen die Beschreibbarkeit paralleler Abläufe sowie temporaler Aspekte. Während Nebenläufigkeit und Verteilung im Information System Engineering meistens konzeptuelle Aspekte darstellen, werden diese Aspekte im Software Engineering oft erst während des Systementwurfs näher betrachtet. Im Gegensatz zum Software Engineering eignen sich die Beschreibungsformalismen nicht zur Darstellung komplexer Algorithmen. Bei der Entwicklung werden innerhalb der Analyse und des Entwurfs oft früh formale Sprachen eingesetzt, die es erlauben, Modelle durch Simulation validieren zu können.

Im Knowledge Engineering werden im Gegensatz zu den anderen Gebieten das Wissen über den konkreten Anwendungsbereich und das Wissen über den Ablauf der Problemlösung, die sog. Problemlösungsmethode, klar getrennt. Die Problemlösungsmethode wird generisch, d.h. unabhängig vom konkreten Anwendungsbereich formuliert. Dies soll zur Wiederverwendung sowohl von Problemlösungsmethoden als auch von Teilen des anwendungsbereichspezifischen Wissens beitragen. Zur Modellierung der Struktur der Daten werden die Möglichkeiten des Information System Engineering noch um geeignete Primitive zur Beschreibung komplexer axiomatischer Zusammenhänge (Regeln und Konsistenzbedingungen) ergänzt. Im Gegensatz zu objektorientierten Methoden des Software Engineering sind Problemlösungsmethoden insbesondere für Probleme, bei denen das Resultat vom Zusammenspiel vieler Objekte abhängt, und für Probleme hoher kombinatorischer Komplexität geeignet. Sie eignen sich deshalb insbesondere zur Modellierung entscheidungsunterstützender Systeme. Allerdings eignen sich die Beschreibungsmittel, genau so wie beim Information System Engineering, nicht zur Darstellung komplexer Algorithmen. Verteilungs- und Zeitaspekte spielen hier praktisch keine Rolle. Ebenso wie im Information System Engineering werden bei der Entwicklung frühzeitig formale Beschreibungsmittel eingesetzt.

Bei näherer Betrachtung der in den unterschiedlichen Gebieten entwickelten Methoden stellt sich allerdings auch heraus, daß die o.g. strikte Trennung der Gebiete nicht gerechtfertigt ist. Es existieren sehr viele Methoden, die jeweils mit leichten Variationen in den unterschiedlichen Methodiken verwendet werden. Allen Entwicklungsmethodiken gemeinsam ist die sukzessive Entwicklung unterschiedlicher Abstraktions- und Formalisierungsebenen, die den Übergang von der rein informalen Beschreibung der Funktion und der Anforderungen an das System bis hin zum realisierten System in mehrere Schritte unterteilen. Beim Information Systems Engineering und beim Knowledge Engineering werden darin auch frühzeitig formale Beschreibungsformalismen angewandt, die die Validierung durch Ausführung bzw. den Zugang für formale Verfahren ermöglichen. Gegenüber rein linearen Vorgehensweisen bei der Entwicklung, wie z.B. durch das klassische Wasserfallmodell beschrieben, scheint es allgemeiner Konsens zu sein, durch Methoden des Prototyping frühzeitig Rückkopplung der erzeugten Modelle gegenüber den tatsächlichen Anforderungen zu gewinnen. Allerdings herrschen in der betrieblichen Praxis meist lineare Vorgehensweisen vor, da z.B. die Überprüfbarkeit von Meilensteinen vertraglichen Vereinbarungen entgegen kommt. Bei den einzelnen Beschreibungsformalismen nähern sich insbesondere die Sprachen zur Beschreibung statischer Strukturen weitgehend aneinander an. Diese Formalismen bieten meist die Möglichkeiten, wie sie für semantische Datenmodelle entwickelt wurden. Vererbungsstrukturen und unterschiedliche Beziehungstypen zwischen Klassen (Entitytypen) sind hier die Regel. Bei der Beschreibung der Dynamik eines Systems werden in allen Gebieten hierarchische Beschreibungsformalismen verwendet, um die Gesamtkomplexität eines Systems besser beherrschbar zu machen. Zur Beschreibung der Dynamik werden im Knowledge Engineering oft Methoden aus dem Software Engineering eingesetzt, die an die Notwendigkeiten des Knowledge Engineering adaptiert wurden.

Es ist klar, daß innerhalb der Arbeitsgruppe nur bestimmte Aspekte der unterschiedlichen Methodiken verglichen werden konnten. So blieben z.B. die Entwicklung von Benutzungsoberflächen, das gesamte Thema Projektmanagement, weitergehende formale Ansätze, wie z.B. Verifikationsmethoden, gänzlich unbetrachtet. Die Einbeziehung aller möglicher Aspekte war aus Zeit- und Aufwandsgründen nicht möglich. So wurden die Betrachtungen auf den jeweiligen Kern einer Entwicklungsmethodik eingeschränkt.

4 Schluß

Jedes der drei Gebiete füllt seine speziellen Anforderungen mit entsprechenden Entwicklungsmethoden und Beschreibungsformalismen aus. Es gibt jedoch sehr viele überlappende Bereiche und Methoden. Somit können alle drei Gebiete von einem Erfahrungsaustausch untereinander profitieren. Die Aufgaben, die Softwaresysteme in Zukunft lösen werden, werden, auch gefördert durch Vernetzung und Verteilung, immer komplexer. Es ist deshalb zu erwarten, daß solche komplexen Gesamtsysteme Anteile enthalten werden, die am besten mit Methoden des Software Engineering entwickelt werden. Andere Anteile werden besser den Methoden des Information Engineering zugänglich sein und wiederum andere Anteile werden wissensbasierte Charakterzüge aufweisen. Es sollten deshalb in Zukunft diese einzelnen Bereiche nicht mehr nur isoliert betrachtet werden. Statt dessen sollten die Anforderungen, die bei der Realisierung von solchen komplexen Systemen an das Zusammenspiel dieser Entwicklungs–methoden gestellt werden auch bei der weiteren Entwicklung der Methodiken berücksichtigt werden. Dies bezieht sich allerdings nicht nur auf die inhaltlichen Aspekte, sondern es müssen auch geeignete Organisationsformen geschaffen werden, um den Erfahrungsaustausch zwischen den unterschiedlichen Gruppen aus den unterschiedlichen Gebieten zu fördern. Mit dieser Arbeitsgruppe wurde dazu ein erster Schritt getan. Ferner sollten in Organisationen, wie z.B. der GI, die strikte Abgrenzung und hierarchische Strukturierung der Fachgebiete durch eine flexiblere Struktur überwunden werden, um Kommunikation und Kooperation zwischen den unterschiedlichen Gruppierungen zu fördern. Darüber hinaus sollten auch innerhalb der Lehre in den einzelnen Gebieten Aspekte aus den anderen Gebieten sehr viel stärker einfließen, um frühzeitig ein Verständnis für die jeweils andere Fachrichtung und für die Integration der Ansätze zu etablieren.

Danksagung

Dieser Beitrag basiert zu wesentlichen Teilen auf der Langfassung des Abschlußberichtes der GI-Arbeitsgruppe. Unser spezieller Dank gilt deshalb den anderen Autoren der Langfassung: B. Dellen, F. Maurer, A. Oberweis, G. Pews, G. Zimmermann, W. Stein. Die Autoren danken allen weiteren (zeitweiligen) Mitgliedern der Arbeitsgruppe für ihre konstruktive Mitarbeit: H.J. Cleef, D. Landes, M. Leppert, P. Löhr-Richter, S. Neubert, H.J. Ott, Th. Pirlein, M. Riebisch, G. Schöpke, A. Schwanke, B. Thelen und H. Voß.

Literatur

[APS+97] J. Angele, R. Perkuhn, R. Studer, A. Oberweis, G. Zimmermann, B. Dellen, F. Maurer, G. Pews, W. Stein: Abschlußbericht der GI-Arbeitsgruppe "Vergleichende Analyse von Problemstellungen und Lösungsansätzen in den Fachgebieten Information Systems Engineering, Software Engineering und Knowledge Engineering". In: *EMISA-Forum*, Heft 2, Mitteilungen der GI Fachgruppe 2.5.2, Entwicklungsmethoden für Informationssysteme und deren Anwendung, 1997.

[AFS96] J. Angele, D. Fensel und R. Studer: Domain and Task Modeling in MIKE. In: A. Sutcliffe, D. Benyon, F. van Assche (Eds.): *Domain Knowledge for Interactive System Design, Proceedings of IFIP WG 8.1/13.2 Joint Working Conference*, Geneva, May 1996, Chapman & Hall, 1996.

[ALO+93] J. Angele, D. Landes, A. Oberweis und R. Studer: Vorgehensmodelle und Methoden zur Systementwicklung. In: H. Reichel (ed.): *Informatik - Wirtschaft - Gesellschaft, Proc. 23. GI-Jahrestagung*, Dresden, September 1993, Informatik aktuell, Springer, Berlin, 1993.

[AnS93] J. Angele und R. Studer (eds.): Arbeitsunterlagen zum Workshop *Querbezüge des Knowledge Engineering zu Methoden des Software*

Engineering und der Entwicklung von Informationssystemen, 2. Deutsche Tagung Expertensysteme, Hamburg, Februar 1993.

[DKS93] J.-M. David, J.-P. Krivine, R. Simmons (eds.): Second Generation Expert Systems, Springer, 1993.

[DMP97] B. Dellen, F. Maurer, G. Pews: Knowledge Based Techniques to Increase the Flexibility of Workflow Management, to appear *in Data & Knowledge Engineering Journal*, North-Holland, 1997.

[FAS97] D. Fensel, J. Angele, and R. Studer: The Knowledge Acquisition and Representation Language KARL. To appear in*: IEEE Transactions on Knowledge and Data Engineering*, 1997.

[Hüs95] T. Hüsener: Objektorientierter Entwurf von nebenläufigen, verteilten und echtzeitfähigen Software-Systemen, Dissertation, Spektrum Akademischer Verlag Heidelberg, 1995

[Kar90] D. Karagiannis (ed.*): Information Systems and Artificial Intelligence: Integration Aspects*. Lecture Notes in Computer Science 474, Springer, Berlin, 1990.

[Kri97] Kriterienkatalog für den Vergleich von Softwareentwicklungs–methodiken.
ftp://aifbmozart.aifb.uni-karlsruhe.de/pub/mike/Fragebogen.ps.Z

[KuS94] J. Kunze und H. Stoyan (eds.): *KI-94 Workshops, 18. Deutsche Jahrestagung für Künstliche Intelligenz*, Saarbrücken, September 1994, Gesellschaft für Informatik e.V.

[LuM94] K. von Luck und H. Marburger (eds.*): Management and Processing of Complex Data Structures - Proc. 3rd Workshop Information Systems and Artificial Intelligence*, Lecture Notes in Computer Science 777, Springer, Berlin, 1994.

[Neu93] S. Neubert: Model Construction in MIKE (Model-Based and Incremental Knowledge Engineering). In: N. Aussenac, G. Boy, B. Gaines, M. Linsert, J.-G. Ganascia, Y. Kodratoff (eds.): *Knowledge Acquisition for Knowledge-Based Systems, Proceedings of the 7th European Workshop (EKAW'93),* Toulouse, France, September 6-10, 1993), Lecture Notes in AI no 723, Springer-Verlag, Berlin, 1993.

[NOS92] T. Németh, A. Oberweis, F. Schönthaler, W. Stucky: INCOME: Arbeitsplatz für den Programmentwurf interaktiver betrieblicher Informationssysteme. Forschungsbericht 251, Institut für Angewandte Informatik und Formale Beschreibungsverfahren, Universität Karlsruhe, 1992.

[INC93] INCOME User Manuals: INCOME/Designer, INCOME/Dictionary, INCOME/ Generator, PROMATIS Informatik, Karlsbad, 1993.

[OSV82] T.W. Olle, H.G. Sol und A.A. Verrijn-Stuart (eds.): *Information Systems Design Methodologies: A Comparative Review*. North-Holland, 1982.

[OSS94] A. Oberweis, G. Scherrer und W. Stucky: INCOME/STAR: Methodology and tools for the development of distributed information systems. Information Systems, Vol. 19(8) 1994, S. 641-658

[SoK93] A. Solvberg und D.C. Kung: Information Systems Engineering, Springer, 1993.

[Som92] I. Sommerville: Software Engineering. Addison-Wesley, 1992.

[Ste95] W. Stein: Objektorientierte Analyse für nebenläufige Systeme, Dissertation, BI-Verlag Mannheim 1995

[Stu92] R. Studer (ed.): *Informationssysteme und Künstliche Intelligenz - Modellierung*. Informatik Fachberichte 303, Springer, Berlin, 1992.

Wiederverwendung von Software-Elementen: Das Java Repository

Peter Buxmann, Frank Rose, Wolfgang König

Zusammenfassung

Die weltweite Entwicklung von Java-Applets und -Klassen eröffnet ein enormes Potential für die Wiederverwendung dieser Software-Elemente mit dem Internet als einer großen „Klassenbibliothek". Das Hauptproblem der Nutzung von Klassenbibliotheken besteht häufig darin, die richtigen Klassen für bestimmte Problemstellungen zu finden. Zur Unterstützung der Suche, Evaluation und anschließendem Austausch von Java-Elementen wurde das Java Repository (http://java.wiwi.uni-frankfurt.de) entwickelt, das in diesem Beitrag vor dem Hintergrund der Wiederverwendung vorgestellt wird.

Abstract

The emergence of the programming language Java offers new opportunities for the reuse of Java software elements. The Internet can serve as a class library. One main problem that arises in the use of software repositories is usually the search for an appropriate class or problem solution. The Java Repository (http://java.wiwi.uni-frankfurt.de) has been developed in order to support search, evaluation, and the exchange of Java software elements over the Internet.

1 Software-Entwicklung im Zeitalter des Internet

Mit der Entwicklung der Programmiersprache Java [Gosling 95] wurde das Internet um die Möglichkeit erweitert, weltweit Anwendungen und Dienste zur Verfügung zu stellen. Seit ihrer Einführung Mitte 1995 nahm die Verbreitung der Sprache ständig zu. Es entstehen weltweit User-Gruppen, und Gremien beschäftigen sich mit der Standardisierung der Sprache. Während Java zu Beginn überwiegend zur graphischen Gestaltung von Webseiten genutzt wurde, hat sich inzwischen die Situation grundlegend verändert. Unternehmen entwickeln mit Java Individual- und Standardanwendungen, prominente Beispiele sind die Software für den Internet-Auftritt der Bank 24 sowie das Office-Paket der Firma Corel.

Die weltweite Entwicklung von Java-Applets und -Klassen (Java-Elemente) und deren Verfügbarkeit im World Wide Web eröffnet ein enormes Potential für die Wiederverwendung dieser Elemente. Das Internet kann hierbei als eine große "Klassenbibliothek" angesehen werden. Das Hauptproblem der Nutzung von Klassenbibliotheken besteht meist darin, die richtigen Klassen für bestimmte Problemstellungen zu finden. Zur Unterstützung der Suche, Evaluation und anschließendem Austausch von Java-Elementen wurde das Java Repository (http://java.wiwi.uni-frankfurt.de) entwickelt. Das Ziel dieses Beitrages besteht darin, diesen Intermediär für Java-Elemente vorzustellen sowie Möglichkeiten und Grenzen der Wiederverwendung vor dem Hintergrund der „Klassenbibliothek Internet“ zu diskutieren.

Das Java Repository ist das Konzept und der Prototyp eines Software-Repository im Internet, von denen es in Zukunft weitere geben wird, die sich auf ein bestimmtes Anwendungsgebiet oder auf eine bestimmte Programmiersprache spezialisieren werden.

2 Wiederverwendung von Java Software-Elementen

Unter Wirtschaftlichkeitsaspekten ist die Wiederverwendung von Software-Elementen seit langer Zeit eine wichtige Anforderung an die Software-Entwicklung. Man versteht darunter allgemein das Benutzen bereits früher erstellter Software-Elemente für die Entwicklung neuer Software [Endres 88, 86]. Die Vorteile der Wiederverwendung liegen auf der Hand, so brauchen bereits verfügbare Software-Elemente nicht neu entwickelt und getestet werden [Sneed 87, 85-86] ("das Rad nicht jedes Mal neu erfinden"). Ermöglicht wird diese Wiederverwendung dadurch, daß

- Software keinem Verschleiß unterliegt,
- die Kopierkosten vernachlässigbar gering sind, und
- viele der durch Software gelösten Problemstellungen sich auf die gleichen Lösungsstrategien zurückführen lassen [Heß 93, 8-9].

Häufig wird das Thema der Wiederverwendung in engem Zusammenhang mit objektorientierter Programmierung genannt. Problematisch wird diese Argumentationskette immer dann, wenn eine direkte Kausalität zwischen Objektorientierung und Wiederverwendbarkeit hergestellt wird. So erwecken manche Autoren den Eindruck, als wenn gerade die Objektorientierung es erst ermöglicht hätte, daß Software-Elemente in anderen Anwendungen wiederverwendet werden. Diese Argumentation greift zu kurz, da bereits in der traditionellen Programmierung Bibliotheken existiert haben, die wiederverwendbare Funktionen enthielten.

Ein wesentlicher Vorteil des objektorientierten Konzeptes in bezug auf die Wiederverwendung besteht in den Möglichkeiten der Vererbung, des Polymorphismus sowie der Verkapselung von Daten und Methoden [Meyer 87]. Dadurch läßt sich ein höherer Abstraktionsgrad und eine eindeutigere Trennung von Nutzung und Implementierung der Klassen erreichen. Insbesondere in Verbindung mit der Möglichkeit zur Modifikation einzelner Teilfunktionalitäten auf Subklassenebene sind Objektklassen einfacher ohne umfangreiche Code-Redundanzen wiederzuverwenden [König/Wolf 1993].

Diese Möglichkeiten der Wiederverwendung stehen natürlich auch Java-Elementen offen. Von noch größerer Bedeutung ist jedoch die weltweite Verfügbarkeit von Problemlösungen im World Wide Web. Mittlerweile finden sich im World Wide Web Informationen zu allen denkbaren Themengebieten, und die Anzahl der verfügbaren Informationen steigt weiterhin exponentiell. Zukünftig wird es dort wahrscheinlich auch fast alle denkbaren kommerziellen Services geben sowie - so unsere Vermutung - Java-Elemente zu einer Vielzahl von Anwendungen und Problemstellungen. Die Ausnutzung dieses Potentials stellt eine große Chance und zugleich Herausforderung für die Zukunft dar und kann zu einer grundlegenden Änderung der zukünftigen Softwareproduktion führen.

Wie bei den meisten Bibliotheken, stellt sich stets das Problem, den richtigen Softwarebaustein für eine gegebene Problemstellung zu finden. Im Internet wird dieses Problem trotz leistungsfähiger Suchmaschinen vermutlich mit der Anzahl vorhandener Java-Problemlösungen sowie der insgesamt wachsenden Menge an Informationen eher zu- als abnehmen. Als ein Service zur Unterstützung der Suche und Auswahl von Problemlösungen wurde das Java Repository entwickelt.

3 Das Java Repository

Das Java Repository bietet Anbietern und Programmierern die Möglichkeit, ihre Software im Java Repository zu registrieren und sie so unter den Nutzern des Repository bekanntzumachen. Das Java Repository (http://java.wiwi.uni-frankfurt.de) ist damit ein Intermediär für Klassen und Applets, die über das World Wide Web verfügbar sind [Buxmann 97]. Hat ein solcher Intermediär einen vollkommenen Marktüberblick, d.h. kann er alle relevanten am Markt verfügbaren Software-Elemente anbieten, so müssen die Nachfrager nur noch mit dem Intermediär und nicht mehr mit allen Anbietern Kontakt aufnehmen.

Eine wesentliche Voraussetzung für eine systematische Wiederverwendung von Software-Elementen ist eine Unterstützung bei der Suche nach geeigneten Bausteinen [Heß 93, 16-19]. Diese Suche wird im *Java Repository* durch zwei Sichten auf die registrierten Einträge unterstützt: Die programmiertechnische Sicht fokussiert mit den verwendeten Methoden und Techniken die Belange des Programmierers, die anwendungsbezogene Sicht ist demgegenüber stärker auf den Anwender ausgerichtet. Das Browsen innerhalb jeder Sicht wird durch vielfältige Sortier- und

Suchmöglichkeiten unterstützt. Durch Schlagwortsuche wird zusätzlich ein direkter Zugriff auf entsprechende Java-Elemente ermöglicht.

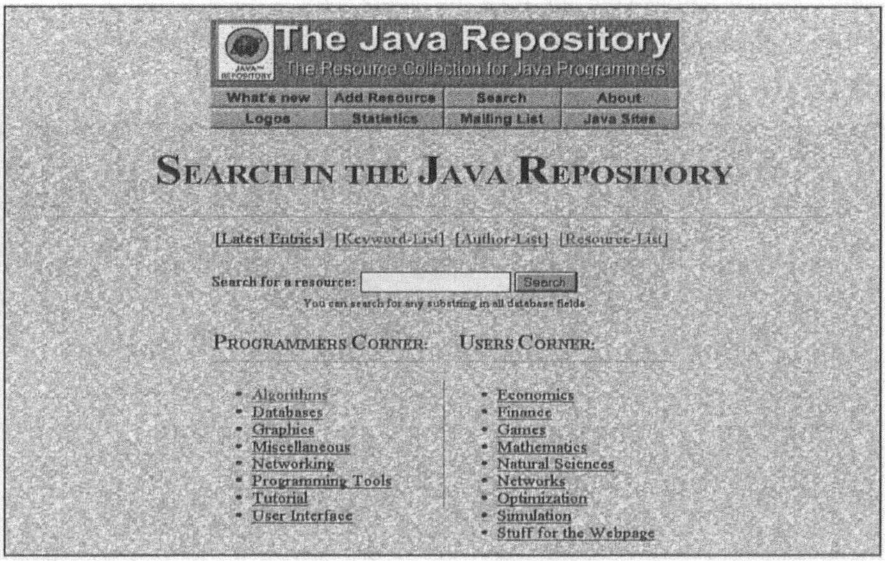

Abb. 1: Suche im Java Repository über Kategorien und direkte Schlagwortsuche

Sind geeignete Java-Klassen oder Applets gefunden, so müssen sie auf ihre Einsetzbarkeit in der konkreten Problemstellung untersucht werden. Die vom Java Repository bereitgestellte Beschreibung der Ressourcen sowie das Rating und die Kommentierung durch andere Anwender leisten einen Beitrag zur Beurteilung der Bausteine. Alle Nutzer des Repository haben die Möglichkeit, registrierte Software-Elemente nach einem standardisierten Verfahren zu bewerten und zu kommentieren. Die so gesammelten Informationen werden zusammen mit der Software präsentiert und können dem Nachfrager bei der Auswahl geeigneter Software-Elemente Hilfestellung geben.

Das Ziel der Bewertungs- und Kommentarmöglichkeit ist es, dem potentiellen Nutzer einer Software bereits vor deren Einsatz oder Erwerb Anhaltspunkte über die Qualität zu geben. Kritisch ist dabei zu beurteilen, daß die Bewertungen von allen

Nutzern, gleich welcher Qualifikation oder Intention, abgegeben werden können und so leicht ein subjektives Bild entstehen kann.[1] Die Kommentierung und Weiterleitung der Kommentare an die Anbieter der im Repository verfügbaren Ressourcen dagegen stellte sich als eine geeignete Möglichkeit heraus, den Kontakt zwischen Anbietern und Nachfragern herzustellen und dem Programmierer wertvolle Hinweise für weitere Entwicklungen zu geben.

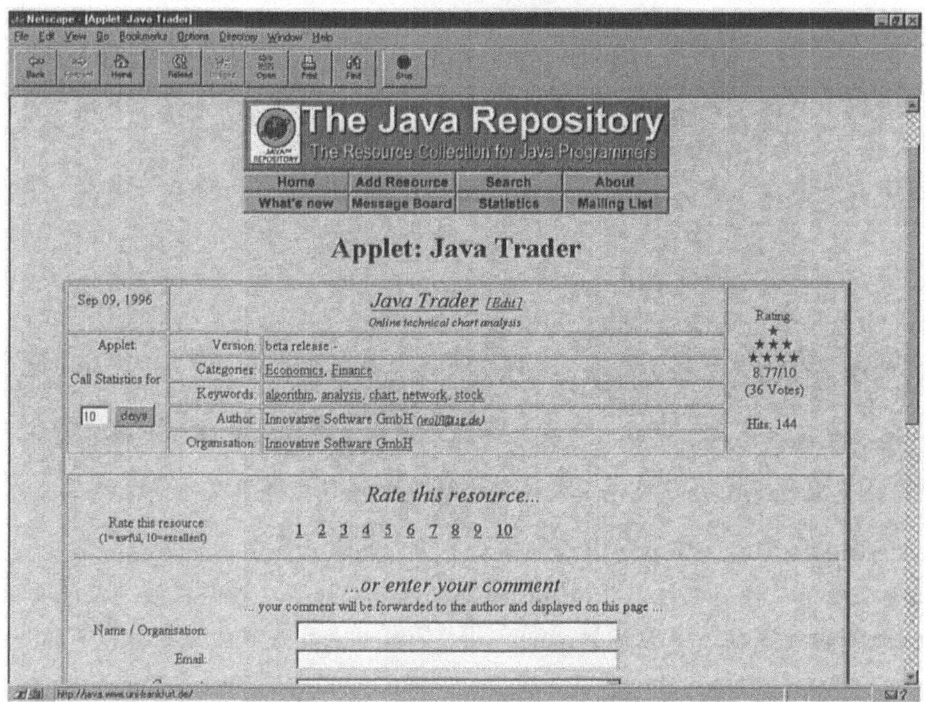

Abb. 2: Darstellung der Bewertung und Kommentierung eines Applet im Java Repository

Das Java Repository wurde realisiert auf einer Sun Ultra Sparc 2, die Anwendung selbst wurde auf Basis einer Oracle-Datenbank mit Webserver entwickelt (Oracle RDBMS 7.3 mit Webserver 2.1, vgl. die Darstellung der technischen Infrastruktur

[1] Um grobe Manipulationen bei der Bewertung zu vermeiden, können einzelne Ressourcen in einem bestimmten Zeitraum nicht mehrfach von der gleichen IP-Adresse aus bewertet werden.

in Abbildung 3). In der Datenbank werden Informationen über den Autor, Adresse sowie zum Teil der Quellcode der Java-Elemente verwaltet. Alle HTML-Seiten zur Navigation und Darstellung werden mit Hilfe von ca. 100 PL/SQL-Prozeduren dynamisch aus der Datenbank generiert.

Zur zukünftigen Unterstützung finanzieller Transaktionen wurde das Ecash-System der Mark-Twain-Bank (http://www.marktwain.com) implementiert [Kalakota 96] [Lynch 96] [Wayner 96]. Hierbei tritt das Java Repository als Intermediär zwischen Anbieter und Nachfrager von/nach Java-Elementen auf. Die Zahlungen werden dabei mit Hilfe von Ecash vom Käufer über das Java Repository direkt an den Anbieter geleitet; dieser zusätzliche Service des *Java Market* ist sowohl für Anbieter als auch Nachfrager kostenlos und optional.

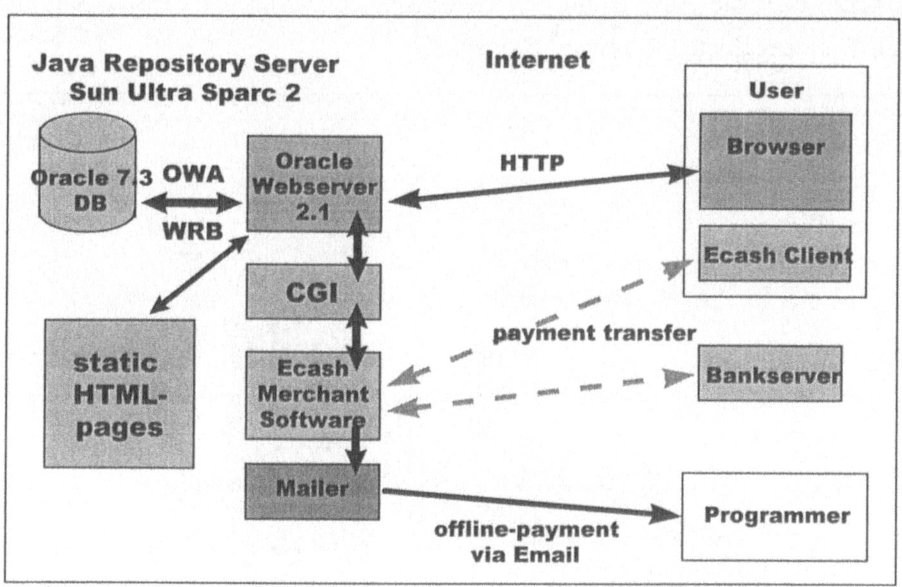

Abb. 3: Technische Realisierung des Java Repository

Das Java Repository besteht seit April 1996 und enthält zur Zeit etwa 750 Java-Elemente für unterschiedliche Problemstellungen. Während anfangs Applets zur Realisierung von Grafikanimationen dominierten, ist in der letzten Zeit ein Trend zu kommerziell orientierten Applets und Applikationen zu beobachten. So enthält das Java Repository beispielsweise 55 Ressourcen in den Kategorien "Finance" bzw. "Economics". Die zunehmende Anzahl der Software-Elemente sowie die Einführung einiger in diesem Abschnitt dargestellten Dienste führte auch zu einer zunehmenden Nutzung des Java Repository. In der folgenden Abbildung ist die auf den Monat aggregierte durchschnittliche Anzahl von Zugriffen pro Tag, d. h. Aufrufen der Hompage von unterschiedlichen Rechnern, sowie die Zahl der Zugriffe auf die Datenbank, d. h. Abfragen und Transaktionen der Datenbank, dargestellt.

Neben dieser meßbaren Nutzung des Repository erhielten wir eine Reihe von Reaktionen zum Service des Repository, die den Service begrüßten und die Implementierung zusätzlicher Dienste, die dann letztlich auch umgesetzt wurden, anregten. Beispiele sind die Einrichtung und Darstellung von Zugriffsstatistiken für jede einzelne Ressource ("Call-Statistik" auf der Beschreibungsseite einer einzelnen Ressource) oder die Auswertung der Häufigkeit der durch die Ressourcen adressierten Themengebiete ("Keyword-Statistik").

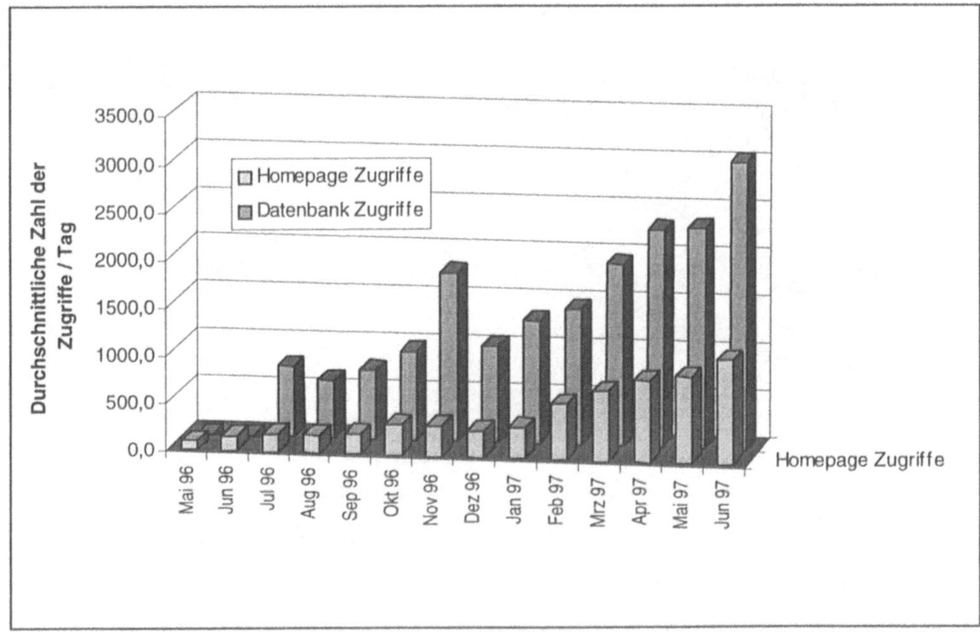

Abb. 4: Durchschnittliche Anzahl von Zugriffen auf die Homepage des Java Repository und Datenbankzugriffe pro Tag

4 Ausblick: Kommerzielle Wiederverwendung von Software im Internet

Die Anzahl der Java-Programmierer und die bereits vorhandenen Java-Elemente bilden ein vielversprechendes Potential für die Wiederverwendung dieser Klassen und Applets. So wie im World Wide Web nahezu alle möglichen Informationen verfügbar sind, wird es zukünftig auch für eine Vielzahl von Problemstellungen Java-Elemente geben. Diese Anwendungen werden - wie die Informationen im World Wide Web - dezentral ohne Vorgabe durch eine zentrale Instanz erstellt und angeboten. Diese dezentrale Struktur ermöglicht erst die Vielfalt der verfügbaren Java-Elemente. Auf der anderen Seite verursacht diese dezentrale Struktur jedoch auch Probleme. In der überwältigenden Vielzahl von Angeboten ist eine gezielte Suche nach Problemlösungen nur schwer und mit hohem Zeitaufwand möglich, zudem besteht keine einheitliche Darstellung und Dokumentation der verfügbaren Software, und es findet keine zertifizierte Qualitätssicherung statt.

Das Java Repository unterstützt im aktuellen Entwicklungsstand den Nutzer - wie im dritten Abschnitt dargestellt - lediglich bei der Suche und vorläufigen Evaluation von Java-Elementen. Um die Potentiale der Wiederverwendung von Java-Elementen aus dem Internet jedoch noch weiter ausschöpfen zu können, sind an ein solches Repository zusätzliche Anforderungen zu stellen, die ähnlich sind wie die Anforderungen an klassische Werkzeuge zur Wiederverwendung [Appelfeller 95], [Heß 93], [Zendler 96] und häufig eine zentrale Organisation benötigen, z.B.

- eine Erweiterung der gespeicherten Informationen auf Ergebnisse der Analyse- und Entwurfsphase, die auch semantische Aspekte sowie Beziehungstypen zwischen den Elementen enthält, wie etwa uses, used-by oder part-of (einschließlich der Nutzung von Frameworks) [Booch 94] [Rumbaugh 91],
- ein von zentraler Stelle durchgeführter Source-Code Review, der einzelne im Repository registrierten Software-Elemente nach fest definierten Qualitätsmaßstäben bewertet,

- eine Dokumentation der Schnittstellen aller Elemente, anhand derer auch ohne Kenntnis der Implementierung über die Wiederverwendbarkeit entschieden werden kann,
- die rechtliche Problematik, den Entwicklern die ihnen zustehenden Erträge zu sichern, ist noch nicht gelöst, auch wenn bereits Lösungsansätze exisitieren [Endres 92],
- Import- und Exportschnittstellen und Integration mit Case-Tools sowie
- umfangreiche Reports und Statistiken zur durchgeführten Wiederverwendung der Softwarelemente.

Die Liste der Anforderungen zeigt, daß viele dieser Maßnahmen zentral geplant und durchgeführt werden müssen, um eine Wiederverwendung von Software-Elementen aus dem Internet wie mit einem unternehmensinternen Software-Repository zu unterstützen.[2] Dabei haben wir die Erfahrungen gemacht, daß diese Aufgaben nicht immer dezentralisiert werden können. So wurde in der Anfangsphase versucht, einen zentralen Dokumentationsstandard auf der Basis des objektorientierten Entwicklungswerkzeugs OEW für Java[3] einzuführen. Die Entwickler der Java-Elemente wurden aufgefordert, neben der „normalen" Registrierung auch eine Dokumentation ihrer Klassen mit diesem Werkzeug zu erzeugen und bereitzustellen. Die Erfahrung zeigte jedoch, daß die meisten Entwickler nicht bereit waren, diesen Aufwand in Kauf zu nehmen, solange sie dafür nicht einen direkten zusätzlichen Nutzen oder eine Entschädigung sahen. Bei dem nicht-kommerziellen Betrieb des Java Repository war der Aufwand einer Nachdokumentation der Software

---

[2] Als weitere Möglichkeit einer *unternehmensübergreifenden* Wiederverwendung diskutiert [Heß 93, 190-198] Shareware Software, die Etablierung von Software Communities sowie Kooperationen zwischen Unternehmen.

[3] OEW für Java ist eine Entwicklungsumgebung der Firma Innovative Software GmbH (http://www.isg.de).

für den Betrieb des Repository nicht zentral durchführbar, so daß die Dokumentation für den Anbieter optional ist.

Einen möglichen Ausweg aus diesem Dilemma stellt die Kommerzialisierung des Betriebes eines solchen Repository im Internet dar. Ein bei kommerziellen Intermediären üblicher Ansatz ist, daß dem Intermediär ein Prozentsatz des zwischen Anbieter und Nachfrager gehandelten Umsatz-Volumens zusteht. Dieser Betrag könnte für die Finanzierung weiterer zentraler Dienste des Intermediärs eingesetzt werden. Diese Dienste könnten wiederum zu einer Steigerung der Attraktivität für den Nutzer führen und dadurch weitere Nachfrager anziehen, was wiederum einen Anreiz für Entwickler darstellt, ein verbessertes Angebot bereitzustellen.

Obwohl diese wünschenswerten weiteren Funktionalitäten heute noch nicht zur Verfügung stehen, erhielten wir eine Vielzahl von Reaktionen, die uns bereits auf dem heutigen Status Quo von der nutzbringenden Wiederverwendung berichteten. Dieses positive Feedback spiegelt sich auch in den stetig zunehmenden Nutzungszahlen des Java Repository wider, die sich, wie Abbildung 4 zeigte, von Mai 1996 bis heute etwa vervierfacht hat.

Literaturverzeichnis

[Appelfeller 95] Appelfeller, W.: Wiederverwendung im objektorientierten Softwareentwicklungsprozeß, dargestellt am Beispiel der Entwicklung eines Lagerlogistiksystems, Frankfurt 1995

[Bailey 96] Bailey, J.: The Emergence of Electronic Market Intermediaries, in: Proceedings of the 17th International Conference on Information Systems, Cleveland 1996, S. 391-399

[Boehm 81] Boehm, B.: Wirtschaftliche Software-Produktion, Wiesbaden 1981

[Booch 94] Booch, G.: Objekt-Oriented Analysis and Design with Applications, Redwood 1994

[Brands 95] Brands, S.: Electronic Cash on the Internet, Proceedings of the Internet Society 1995 Symposium on Network and Distributed Systems Security, San Diego, California, USA

[Buxmann 97] Buxmann, P.; König, W.; Rose, F.: Aufbau eines elektronischen Handelsplatzes für Java-Applets, in: Krallmann, H. (Hrsg.): Wirtschaftsinformatik 1997, S. 35-47

[Endres 88] Endres, A.: Software-Wiederverwendung: Ziele, Wege und Erfahrungen, in: Informatik-Spektrum 11 (1988), S. 85-95

[Endres 92] Endres, A.: Der rechtliche Schutz von Software: Aktuelle Fragen und Probleme, in: Informatik-Spektrum 15 (1992), S. 89-100

[Frakes 95] Frakes, W.; Fox, C.: Sixteen Questions about Software Reuse, in: Communications of the ACM, 6 (1995), S. 75-87

[Gosling 95] Gosling, J.; McGilton, H.: The Java Language Environment - A White Paper, Sun Microsystems Computer Company, Oktober 1995

[Heß 93] Heß, H.: Wiederverwendung von Software, Wiesbaden 1993

[Jones 94] Jones, T.: Economics of Sotware Reuse, in: IEEE Computer, 1994

[Kalakota 96] Kalakota, R; Whinston, A.: Frontiers of Electronic Commerce, Addison Wesley, 1996

[König 93] König, W.; Wolf, S.: Objektorientierte Software-Entwicklung - Anforderungen an das Informationsmanagement, in: Scheer, A.-W.: Handbuch Informationsmanagement, Wiesbaden, 1993

[Lynch 96] Lynch, D.; Lundquist, L.: Digital Money, John Wiley & Sons Inc., New York, 1996

[Meyer 87] Meyer, B.: Reusability: The Case for Object-oriented Design, IEEE Software, 4 (1987), S. 50-64

[Rumbough 91] Rumbaugh, J. et al.: Object-Oriented Modeling and Design, Prentice Hall, Englewood Cliffs 1991

[Sarkar 96] Sarkar, M.; Butler, B.; Steinfeld, C.: Intermediaries and Cyber-mediaries: A Continuing Role for Mediating Players in the Electronic Marketplace, (http://shum.cc.huji.ac.il/jcmc/vol1/issue3/sarkar.html, March, 27th 1997)

[Sneed 87] Sneed, H. M.: Software Management, Köln 1987

[Wayner 96] Wayner, P.: Digital Cash: Commerce on the Net, Boston et al. 1996

[Zendler 96] Zendler, A.: Kommerzielle Werkzeuge zur Administration von wiederverwendbaren Software-Dokumenten, in: Wirtschaftsinformatik 2 (1996), S. 147-159.

Eine Studie zur Erzielung kürzerer EDV-Projektlaufzeiten in einem Versicherungsunternehmen

Helmut Häck

Abstract

Die EDV-Projektlaufzeit ist bedeutsam für den wirtschaftlichen Erfolg der Software-Entwicklung. Dieser Artikel zeigt wie das Ziel "Verkürzung von EDV-Projektlaufzeiten" erreicht wird. Dabei ist es nicht das Ziel der Studie, konkrete, ausgearbeitete und somit sofort in Projekten einsetzbaren Maßnahmen zu finden. Vielmehr liegt der Schwerpunkt in dem "wissenschaftlichen" Finden von Kriterien und dazugehörigen Maßnahmenideen. Das "Wissenschaftliche" findet seine Bestätigung in der Berücksichtigung eines Tools, interner und externer Experten sowie eines Arbeitsteams zur Lösung dieser Aufgabe.

1 Der Projektauftrag

Das Management der Colonia Nordstern Versicherungsmanagement AG (CNV) hat eine Untersuchung in Auftrag gegeben mit der Aufgabenstellung, Maßnahmen für eine Verkürzung von EDV-Projektlaufzeiten zu finden. Der diesbezüglichen Konkretisierung, die Laufzeit nach Abschluß des groben Fachkonzeptes auf maximal 18 Monate zu reduzieren, wurde während der Untersuchung weniger Bedeutung zugemessen, da das grundlegende Finden von geeigneten Maßnahmen im Vordergrund stand. Die Brisanz des Managementauftrages verdeutlicht folgende Erläuterung der Wichtigkeit der Verkürzung von EDV-Projektlaufzeiten.

Die **EDV-Projektlaufzeit** ist eine besonders kritische Einflußgröße auf den wirtschaftlichen Erfolg der entsprechenden Software-Entwicklung und trägt somit auch langfristig zum Unternehmenserfolg bei. Mit EDV-Projektlaufzeit ist allgemein die Zeit gemeint, die ein EDV-Projekt bis zu seiner Fertigstellung benötigt. Sie ergibt sich aus der Differenz zwischen End- und Starttermin des Projektes. Für eine Verkürzung der Laufzeit sprechen hauptsächlich zwei Gründe:

Erstens, die schnellen **Veränderungen**. Dafür gibt es mehrere Verursacher:

- Schleichender Funktionszuwachs (**requirements creep**):

  Die zu Beginn der Entwicklung bestehenden Anforderungen an das Softwareprodukt ändern bzw. erweitern sich während der Entwicklungsdauer relativ schnell und häufig. So wird von einem Funktionszuwachs von mindestens einem Prozent pro Monat Projektlaufzeit ausgegangen. Dies bedeutet bei einer Projektlaufzeit von zwei Jahren, daß mit dem Projektende auf jeden Fall 24% mehr realisiert worden ist als anfänglich vorgesehen war.[Jon96]

- Organisatorischer / unternehmerischer Zwang:

  Es ist möglich, daß der betroffene Fachbereich zwischenzeitlich aufgelöst, umstrukturiert oder auf andere Verfahren umgestellt werden soll.

- Gesetzlicher Zwang:

  Beispielsweise durch neu verabschiedete steuerliche Gesetzgebungen, die das zu erstellende Softwareprodukt betreffen.

- Technischer Zwang:

  Mögliche Veränderungen können sich ebenfalls durch eine schnelle Technologie-Entwicklung ergeben. "Die Softwareentwicklung hinkt der technologischen Entwicklung im Bereich der Hardwaretechnologie hinterher, ... Marktanalysten gehen derzeit von einer Verdopplung der Rechenleistung alle 18 Monate aus."[Sch96]

Der zweite Grund für eine Verkürzung der Laufzeit besteht in der Konkurrenz. Hier ist die Zeit ein nicht zu unterschätzender **Wettbewerbsfaktor**. Ein Unternehmen kann nur überleben, wenn es in der Lage ist, ein neues Produkt schnell auf dem

Markt zu präsentieren. Folglich muß das Unternehmen eine Änderungs- bzw. Anpassungsgeschwindigkeit an den Markt besitzen, die größer oder gleich der seiner Konkurrenten ist. Dieser Sachverhalt wird **"time to market"** bezeichnet.

2 Kriterienfindung

2.1 Kriterienfindung durch Simulationen mit Checkpoint

Der erste Ansatz, um Kriterien zu finden, die eine Verkürzung von EDV-Projektlaufzeiten ermöglichen, war die Durchführung von Simulationen mit dem relativ neuen Schätztool Checkpoint. **Kriterien** sind als Einflußmöglichkeiten auf die Projektlaufzeit aufzufassen. Sie führen jedoch erst zur Verkürzung der Projektlaufzeit, wenn entsprechende Maßnahmen zur Verbesserung des jeweiligen Kriteriums gefunden worden sind.

Checkpoint ist ein wissensbasiertes Expertensystem der Fa. SPR (Software Productivity Research), das unter Berücksichtigung von ca. 200 Projekt-Umfeldfaktoren auf der Basis einer umfangreichen Datenbank über derzeit ca. 6.700 Projekte die Planung und Schätzung des Aufwandes von Projekten unterstützt. Mit Checkpoint sind, durch die Eingabe unterschiedlicher Parameter, in geeigneter Weise Simulationen möglich. Deswegen hieß das konkrete Ziel der Simulationen, die effektivsten Attribute (Parameter) bezüglich der Projektdauer am Beispiel eines typischen CNV-Projektes zu ermitteln.

Für die Simulationen bestand folgendes zusammengefaßtes Vorgehen: Als erstes wurde für das Simulationsprojekt die Sensitivitätsanalyse aktiviert, um von Checkpoint die Parameter zu erhalten, die noch bei entsprechender Veränderung eine Laufzeitverkürzung ermöglichen. Mit der **Sensitivitätsanalyse** zeigt Checkpoint zu den Projektzielen Laufzeit, Aufwand, Produktivität und Qualität jeweils die stärksten Parameter. Somit werden die Parameter aufgelistet, die auf das entsprechende Ziel - unabhängig der aktuellen Parameterbewertung - die stärksten Auswirkungen haben. Für die Untersuchung war lediglich das Projektziel Laufzeit von Relevanz.

Nach der Sensitivitätsanalyse sind diese Parameter sukzessive und einzeln um eine Bewertungseinheit verbessert (z.B. von 4 auf 3), in Tabellen dokumentiert und anschließend wieder zurückgesetzt worden. Für die Bewertung der Attribute verwendet Checkpoint eine Skala von eins bis fünf. Die schrittweise Variation der Parameter aus der Sensitivitätsanalyse ergab nach der Zeitverkürzung sortiert folgende Werte.

Die jeweiligen Parameter um 1 erhöht!	ZIELGRÖSSEN													
	Dauer in Tagen		Zeitverkürzung				Aufwand in PM		Anzahl Mitarbeiter		Qualität			
											Fehleranzahl bei Auslieferung		Fehlerentfernungsrate	
Ziel:	Standard	Zeit	Standard		Zeit		Standard	Zeit	Standard	Zeit	Standard	Zeit	Standard	Zeit
PARAMETER			Tage	%	Tage	%								
Grundeinstellung	914	616			298	32,60	852,69	835,13	86	138	1255	1335	78,75	78,55
1) Individual office environment	843	577	71	**7,77**	337	**36,87**	807,71	791,35	86	141	1255	1335	78,75	78,55
2) Development personnel application experience	864	591	50	**5,47**	323	**35,34**	773,39	756,00	68	132	1188	1263	78,58	78,29
3) Office noise and interruption environment	880	597	34	**3,72**	317	**34,68**	828,79	811,57	86	141	1255	1335	78,75	78,55
4) Project organization structure	886	594	28	**3,06**	320	**35,01**	845,32	828,17	86	138	1255	1335	78,75	78,55
5) Product memory utilization restrictions	890	604	24	**2,63**	310	**33,92**	836,52	819,68	86	138	1255	1335	78,75	78,55
6) Product performance/execution speed restrictions	890	604	24	**2,63**	310	**33,92**	836,52	819,68	86	138	1255	1335	78,75	78,55
7) Development personnel tool and method experience	890	604	24	**2,63**	310	**33,92**	837,39	820,34	86	138	1255	1335	78,75	78,55
8) Tools, equipment, and supplies	891	604	23	**2,52**	310	**33,92**	838,43	821,23	86	138	1255	1335	78,75	78,55
9) Functional novelty	895	606	19	**2,08**	308	**33,70**	825,83	807,98	86	145	1186	1260	78,54	78,26
10) New data complexity	898	596	16	**1,75**	318	**34,79**	774,24	758,97	84	138	1161	1240	77,83	77,56
11) New code complexity	903	596	11	**1,20**	318	**34,79**	731,70	714,43	79	130	1051	1127	76,49	76,27
12) New problem complexity	903	596	11	**1,20**	318	**34,79**	731,70	714,43	79	130	1051	1127	76,49	76,27
13) Design automation environment	917	611	-3	**-0,33**	303	**33,15**	815,49	795,33	86	135	1119	1178	77,51	76,98

Tabelle 1: Sensitivitätsparameter einzeln um 1 erhöht

Die wichtigsten Werte gemäß Zielsetzung sind in der Spalte Zeitverkürzung enthalten. Es wird absolut in Tagen und prozentual die mögliche Zeitverkürzung des Projektes aufgrund der entsprechenden Parametermodifizierung für das Ziel Standard und Zeit aufgeführt. Dabei bezieht sich die Zeitverkürzung mit dem Ziel Standard und Zeit jeweils auf die Grundeinstellung mit der Dauer von 914 Tagen. Das Projektziel Standard aus der Grundeinstellung des Simulationsprojektes steht für die Standardeinstellung der Projektziele Zeit, Mitarbeiter und Qualität. Damit jedoch eine bessere Zeitverkürzung erreichbar ist, wurde zusätzlich das Projektziel alternativ auf Zeit gesetzt. Konkret bedeutet dies, das Projekt in der kürzesten Zeit mit mehr Mitarbeitern fertig zu stellen. So verdeutlicht die Tabelle 1, daß mit Hilfe der Zielvariation - Standard durch Zeit ersetzt - bei der Grundeinstellung bereits eine **Zeitverkürzung von 32,60% ohne Parametermodifikation** möglich ist.

Für eine bessere Beurteilung potentieller Auswirkungen und zum Erkennen möglicher "Ausreißer" sind auch die Zielgrößen Aufwand in Personenmonate (PM), Anzahl Mitarbeiter und die Qualität berücksichtigt worden. Ferner ist die Qualität in Fehleranzahl bei Auslieferung und Fehlerentfernungsrate aufgeteilt. Es ist eine große prozentuale Fehlerentfernungsrate anzustreben. Ebenso ist eine geringe Fehleranzahl bei Auslieferung der Software positiv zu bewerten. Das Ablesen der Tabelle 1 ist einfach. Beispiel: mit dem stärksten Parameter 1) kann die Projektlaufzeit mit dem Projektziel Zeit um 36,87% bei gleichbleibender Qualität verkürzt werden, jedoch steigt dann die Anzahl der benötigten Mitarbeiter auf 141 an.

Als nächstes erfolgte die Summierung der einzelnen Parameter. Anschließend wurden die Parameter einzeln und hiernach summenmäßig auf eins verbessert. Sämtliche Simulationen sind in zu der Tabelle 1 analogen Tabellen dokumentiert worden. Deren Vorstellung würde den Rahmen dieses Berichtes “sprengen“.

Mit Hilfe der durchgeführten Simulationen ist deutlich geworden, daß es eine Vielzahl von Möglichkeiten gibt, um die Projektlaufzeit von EDV-Projekten zu verkürzen. Jedoch haben sich einige Parameter hervorgehoben - sind in dem Prozeß der im nächsten Abschnitt beschriebenen Kriterienfindung eingeflossen - wie z.B. die Involvierung der Anwender, die durch den Fachbereich (FB) im Versicherungsunternehmen vertreten werden.

2.2 Kriterienfindung durch das Arbeitsteam

Zum Finden weiterer Kriterien wurde ein Arbeitsteam gebildet. Teilnehmer des **Arbeitsteams** waren neben dem Autor ein Unternehmensberater und zwei Mitarbeiter der Abteilung Projektmanagement und Qualitätssicherung der CNV. Bei der Kriterienfindung mögliche Herkunftsarten waren neben den bereits erwähnten Checkpoint-Simulationen die interne und externe Erfahrung sowie die Literatur. In der anschließenden Tabelle 2 sind die im Arbeitsteam gefundenen 24 Kriterien enthalten.

Unter **interner Erfahrung** ist die Erfahrung der Mitarbeiter zu verstehen, insbesondere diejenige, die aus einem Projektleiter-Workshop ermittelt werden konnte.

Kriterien	interne Erfahrung	externe Erfahrung	Checkpoint (Simulation)	Literatur
Projektziele (Qualitätsanforderungen)	x	x	x	x
Projektnutzen	x	x	x	x
Softwaretechnologie	x		x	x
FB-Beteiligung			x	
Mitarbeiterqualifikation	x	x	x	x
Büro-Ergonomie		x	x	x
Kommunikation	x	x		x
Schnittstellen	x			
Umfeld-Einflüsse	x			x
Projekt-Organisation	x	x		
Test	x	x		
Entwicklungsprozeß	x	x		x
Projektcontrolling	x	x		x
Projektstrukturierung (Parallelisierung)		x		
Projekt-Vorbereitung		x		x
Lernfähigkeit	x	x		x
Kultur		x		x
Motivation		x		x
Bürokratie	x			
PM-Vorgaben	x	x		x
Wettbewerb mit Externen		x		
Ressort-Struktur		x		x
Dokumentation				x
Risiko		x		x

Tabelle 2: Kriterienherkunft

Die **externe Erfahrung** beinhaltet das Wissen von verschiedenen Unternehmensberatungen, welches in Prospekten und Zuschriften vorlag. Als vierte Herkunftsart der Kriterien ist die **Literatur** in der o.a. Tabelle aufgeführt. Es werden viele Kriterien auch in der Literatur genannt bzw. beschrieben.[Bun97] Alle 24 Kriterien wurden CNV-spezifisch definiert.

3 Kriterienauswahl und Maßnahmenvorschläge

3.1 Kriterienauswahl durch das Arbeitsteam

Eine Kriterienauswahl ist erforderlich, da nur eine begrenzte Anzahl von Kriterien mit dazugehörigen Maßnahmen möglichst schnell und effektiv in der CNV zu verwirklichen sind. Die Kriterienauswahl des Arbeitsteams gliederte sich in zwei wesentliche aufeinander aufbauende Schritte. Im ersten Schritt fand die Bewertung aller Kriterien statt, als zweites wurden die Kriterien auf Basis von zwei Selektionen ausgewählt.

3.1.1 Kriterienbewertung

Bevor die gefundenen Kriterien sinnvoll ausgewählt werden konnten, war eine Kriterienbewertung als Entscheidungsgrundlage vorzunehmen. Dazu wurden die Kriterien nach unterschiedlichen Gesichtspunkten (Wertungsfelder) bewertet. Nachfolgend erfolgt eine Aufzählung der **Wertungsfelder** mit der dazugehörigen Bedeutung:

- **Note Istzustand (lfd. Entwicklung):** Ist der heutige Zustand, wie er bei der täglichen Arbeit erlebt wird.
- **Erfolg und Nutzen:** Wie wertvoll ist der Nutzen für die eigene Arbeit zu sehen, wenn auf dem Gebiet des jeweiligen Kriteriums Verbesserungen erfolgen würden.
- **Widerstand (Umsetzbarkeit):** Bewertung der zu erwartenden Schwierigkeiten bei Veränderungen. Es kann z.B. mangelnde Akzeptanz bei den betroffenen Organisationseinheiten entstehen.
- **Aufwand für die Umsetzung:** Bewertung des Aufwands, um Verbesserungen für das Kriterium umzusetzen. Ist im wesentlichen der interne Personalaufwand und enthält nicht die Sachkosten.
- **Sachkosten für die Einführung:** Einschätzung der Kosten für z.B. Tools oder erforderliche Umbauten.
- **Umsetzungszeit:** Schnelligkeit der Wirksamkeit von Verbesserungsmaßnahmen oder anders formuliert: Eintreten des Erfolgszeitpunktes.

Das Arbeitsteam hat sich für nachstehende Kriterienbewertung in den verschiedenen Gesichtspunkten entschieden.

		Gesichtspunkte (Wertungsfelder)					
Kriterien-Nr.	Kriterien	Note Istzustand (lfd. Entwickl.)	Erfolg und Nutzen	Widerstand (Umsetzbarkeit)	Aufwand für Umsetzung	Sachkosten für Einführung	Umsetzungs-zeit
1	Projektziele (Qualitätsanforderungen)	3	1	4	3	1	2
2	Projektnutzen	3	1	1	2	1	2
3	Softwaretechnologie	2	1	3	5	5	4
4	FB-Beteiligung	4	1	4	1	2	2
5	Mitarbeiterqualifikation	2	1	3	4	4	2
6	Büro-Ergonomie	3	1	3	5	5	2
7	Kommunikation	2	1	3	2	2	1
8	Schnittstellen	4	1	2	2	1	2
9	Umfeld-Einflüsse	3	2	5	5	5	5
10	Projekt-Organisation	3	2	4	2	1	2
11	Test	3	1	5	4	3	2
12	Entwicklungsprozeß	3	1	3	4	2	3
13	Projektcontrolling	3	1	4	2	2	3
14	Projektstrukturierung (Parallelisierung)	4	1	2	2	2	2
15	Projekt-Vorbereitung	2	2	2	2	2	2
16	Lernfähigkeit	5	1	4	4	3	4
17	Kultur	4	1	4	2	1	4
18	Motivation	3	1	4	2	1	2
19	Bürokratie	4	2	4	4	1	1
20	PM-Vorgaben	3	1	4	2	1	2
21	Wettbewerb mit Externen	3	3	5	4	4	4
22	Ressort-Struktur	3	1	5	3	2	3
23	Dokumentation	3	3	2	2	2	2
24	Risiko	3	2	2	2	1	2
	Notenerklärung (es sind auch die Noten zwischen 1 und 5 zu vergeben)	1 = sehr guter Zustand 5 = stark verbesserungswürdig	1 = sehr guter Nutzen 5 = kaum Nutzen	1 = geringer Widerstand 5 = sehr hoher Widerstand	1 = geringer Aufwand 5 = sehr hoher Aufwand	1 = geringe Kosten 5 = sehr hoher Kosten	1 = sofort 2 = innerhalb Projektzeitraum 5 = sehr langfristig
	Durchschnittnote	3,13	1,38	3,42	2,92	2,25	2,50

Tabelle 3: Kriterienbewertung des Arbeitsteams

Es fällt auf, daß nach dieser Bewertung Erfolg und Nutzen der Kriterien insgesamt mit sehr gut (Durchschnittsnote 1,38) bewertet worden ist, aber überdurchschnittlicher Widerstand für die Umsetzung (Durchschnittsnote 3,42) erwartet wird. Die Durchschnittsnoten ließen sich ermitteln, in dem für jeden Gesichtspunkt die Summe der einzelnen Kriteriennoten durch die Kriterienanzahl 24 dividiert wurde.

3.1.2 Zwei Selektionen der Tabelle 3

Mit der ersten Selektion sind die Kriterien gesucht worden, die einen schlechten Istzustand besitzen und bei Verbesserungen einen hohen Erfolg und Nutzen versprechen. Dazu wurden die Kriterien mit einem Istzustand von 5 oder 4 und einem Erfolg von 1 oder 2 selektiert. In einer zweiten Selektion wurden die Summen der übrigen Wertungsfelder gebildet, um die Kriterien mit geringem Widerstand, Aufwand und Sachkosten sowie mit schnellem Erfolg zu finden.

Das Arbeitsteam entschied sich für **zwei Ergebnisse**:

1. Kriterien, bei denen eine Übereinstimmung zwischen der ersten und der zweiten Selektion vorhanden ist. Dieses Ergebnis wird im weiteren Verlauf der Arbeit

als **"Arbeitsteam: Übereinstimmungsbewertung"** bezeichnet und beinhaltet die Kriterien: 1, 2, 4, 8, 14, 18, 19, 20.

2. Nach Meinung des Arbeitsteams sind in einem zweiten Ergebnis aufgrund der Wichtigkeit zusätzlich die Kriterien 12, 13, 16 und 17 aufzunehmen. So ist z.B. der Entwicklungsprozeß (Kriterium-Nr. 12) ein lfd. Projekt in der CNV und die Lernfähigkeit (Kriterium-Nr. 16) ein Hauptziel des Entwicklungsprozeßprojektes. Dieses Ergebnis wird im weiteren Verlauf der Arbeit als **"Arbeitsteam: Gesamtbewertung"** bezeichnet und beinhaltet die Kriterien: 1, 2, 4, 8, 12, 13, 14, 16, 17, 18, 19, 20.

Die Bedeutung und Behandlung der Unterscheidung dieser zwei Ergebnisse wird im Kapitel "Endauswahl der Kriterien" verdeutlicht. Doch zunächst waren mögliche Maßnahmen zu den Kriterien zu finden.

3.2 Maßnahmenvorschläge durch das Arbeitsteam

Die in diesem Abschnitt skizzierten zwei Maßnahmen sind nicht als fertige bzw. endgültige Maßnahmen anzusehen, sondern sind erste weiter zu konkretisierende Ideen und dienen als Basis für die im Kapitel 3.3 beschriebenen Expertenbefragungen. Ferner lassen sich nachfolgend aufgeführte Maßnahmen den zuvor ermittelten Ergebniskriterien unmittelbar zuordnen.

3.2.1 Aufbauorganisation

Diese Maßnahme bezieht sich auf das vierte Kriterium der Fachbereichsbeteiligung. Mit der Abbildung 1 wird die bestehende Aufbauorganisation bezüglich der Zusammenarbeit zwischen den Bereichen Datenverarbeitung und Fachbereich (FB) für die Projekte dargestellt.

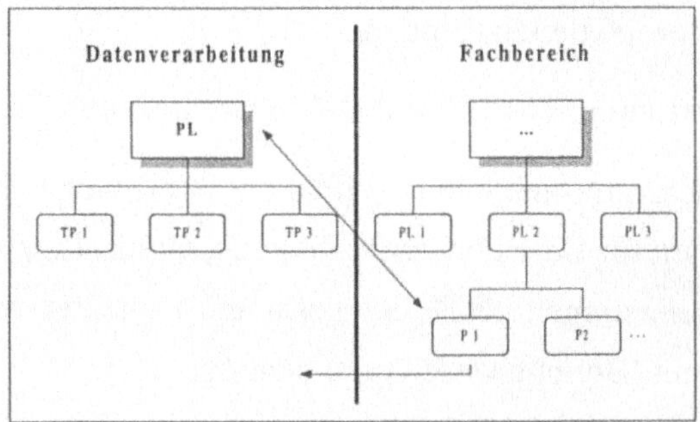

Abb. 1: Bestehende DV-FB-Aufbauorganisation

In der Datenverarbeitung kann ein Projektleiter (PL) mehrere Teilprojekte (TP) führen; im Fachbereich kann ein PL für verschiedene Projekte zuständig sein. Der trennende vertikale Strich zwischen den beiden Bereichen symbolisiert eine schwer zu überwindende "Mauer". Die Pfeile weisen auf typische praktizierte Kommunikationsverbindungen hin. Durch diese "Trennung" ergeben sich für die Projektarbeit folgende **Probleme**:

- Kommunikationsdefizite zwischen DV und FB
- Mangelnde Verfügbarkeit im FB
- Unklare Kompetenzen

Wegen der o.a. Probleme hatte das Arbeitsteam nachfolgende Verbesserung der Aufbauorganisation erarbeitet:

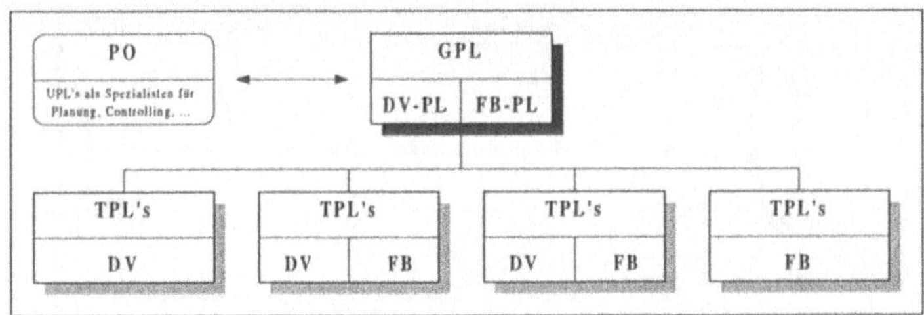

Abb. 2: Verbesserte DV-FB-Aufbauorganisation

Für die Realisierung der in Abbildung 2 skizzierten Aufbauorganisation sind einige Maßnahmen erforderlich:

- Verankerung der Zusammenarbeit DV und FB (siehe Abb. 2)

- Regelmäßige Teilprojektleiter-Runden
- Gleichmäßige Vertretung von FB und DV in Gesamt- und Teilprojektleitung
- Entlastung des Gesamtprojektleiters (GPL) durch effizientes Projektoffice (PO), das aus Spezialisten für Berichtswesen, Planung, Controlling und Qualitätssicherung besteht. Daumenregel: 10 % des gesamten Projektes werden für formales Projektmanagement (Berichtswesen etc.) benötigt.
- Räumliche Anordnung der DV- und FB-Teams
- Ein Plan, ein Bericht, ein Projekt

➪
- Projekte werden homogener
- Effizientere Problemlösungen in operativen Projekten
- Transparente Ressourcenverteilung
- Gruppengrößen der Teilprojekte müssen beachtet werden: einstellige Teamgröße

3.2.2 Projektstrukturierung / Parallelisierung

Die Idee der Projektstrukturierung / Parallelisierung liegt in der Gliederung komplexer Aufgaben in mehrere einfache Teilaufgaben. Dabei sollen die Portionen relativ unabhängig und eigenständig realisiert werden und schließlich wieder zum Gesamtkomplex zusammengefügt werden.

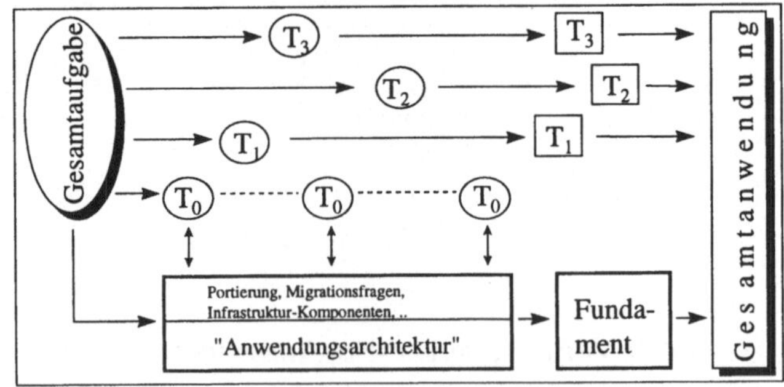

Abb. 3: Aufgabenstrukturierung

Dazu erforderliche **Einzelmaßnahmen** sind:

- Time to market und die weiter unten beschriebene Vorgehensweise als verbindliche Zielvorgaben für die Projekte festlegen
- Auf guter Basis des IAA (Insurance Application Architecture) Datenmodells aufbauen
- Klare Priorisierung der Projektfunktionalitäten (Versionsplan)
- Verbindliche Erstellung einer Projektarchitektur (= Fundament)
- Enge Zusammenarbeit zwischen der PL-Gesamtarchitektur und den Datenmodellverantwortlichen
- Modularisierung von Teilaufgaben (use-cases) durch Identifizierung im Rahmen der Function-Point-Analyse (dabei starke Mitarbeit von FB, Anwendungsmanagement)
- Parallelentwicklung dieser identifizierten use-cases
- Modifikation des Vorgehensmodells
- Entwicklung zu Prototyping, objektorientierter Analyse etc.

Ein Vorschlag zur **Vorgehensweise** ist:

1. Aus Projektzielen und Gesamtarchitektur die Projektarchitektur - in Zusammenarbeit mit Spezialisten für Gesamtarchitektur, Performance, Altsystem und Datenbanken - bilden. Die Projektarchitektur enthält die Kernfunktionalitäten und dient als gemeinsame Basis.

2. Für die Realisierung dieser Projektarchitektur Machbarkeit, Alternativen und Realisierungsplan erarbeiten.

3. Festlegen der Teilaufgabe T_0, die eine spezielle Teilaufgabe in Form eines Prototyps für das Architekturfundament ist.

4. Weitere Teilaufgaben T_1, T_2, ... eigenständig als fachliches Nutzenbündel (use-cases) mit eigenem time to market realisieren.

3.3 Kriterienauswahl und Maßnahmenvorschläge durch die Experten

Die vom Arbeitsteam gefundenen Kriterien und Maßnahmenvorschläge wurden durch interne und externe Experten "abgesichert" bzw. ergänzt. Zu den **externen Experten** gehören drei Unternehmensberatungen mit denen das Versicherungsunternehmen partiell zusammen arbeitet. **Interne Experten** sind Projektleiter der CNV bzw. Teilnehmer der PL-Qualifizierung.

3.3.1 Befragung der externen Experten

Für die in Interviews durchgeführten Befragungen der Unternehmensberatungen wurde ein Interviewleitfaden mit über 30 Fragen erstellt. Damit wurden u.a. folgende Konformitäten der drei Unternehmensberatungen festgestellt: Durchgeführte Studie ist sinnvoll, transparente Ziele, Kickoff-Veranstaltung und regelmäßiges formales Risk-Assessment. Jedoch favorisierten die jeweiligen Unternehmensberatungen bestimmte Aspekte, wie z.B. vertragliche Festlegung von Ziele und Abnahmekriterien sowie Einsatz von Lotus-Notes für eine verbesserte Projektarbeit, Berichtswesen und Kommunikation.

3.3.2 Unternehmensberater-Bewertungen

Die Unternehmensberater haben die Gesichtspunkte, außer Note Istzustand, der Tabelle 3 bewertet. Zur Kriterienauswahl wurden aus diesen Bewertungen die Kriterien ausgesucht, deren Erfolg und Nutzen sowie Erfolgszeitpunkt mit kleiner oder gleich zwei bewertet worden sind. Das sind Kriterien, bei denen - gemäß den Bewertungen - eine Verbesserung mit geeigneten Maßnahmen **erfolgsversprechend und schnell wirkend** beginnt. Mit der o.a. Vorgehensweise ergab sich für die einzelnen Unternehmensberatungen nachstehende Kriterienselektion.

Unternehmensberatung	Kriterien
A	1, 2, 4, 8, 10, 13, 15, 24
B	1, 4, 6, 7, 9, 14, 16, 18, 19, 24
C	1, 2, 4, 7, 10, 13, 14, 15, 18, 20, 24

Diese Kriterienselektion ergab folgende drei Ergebnisse:

Ergebnistyp	Kriterien
Übereinstimmung von allen drei Unternehmensberatungen	1, 4, 24
Übereinstimmung von zwei Unternehmensberatungen	2, 7, 10, 13, 14, 15, 18
Unternehmensberatung: Gesamtbewertung	**1, 4, 24,** 2, 7, 10, 13, 14, 15, 18

Das Ergebnis **"Unternehmensberatung: Gesamtbewertung"** wird im Kapitel "Endauswahl der Kriterien" weiter behandelt. Zunächst steht aber die Befragung der internen Experten (Projektleiter) auf.

3.3.3 Befragung der internen Experten

Zur Befragung der Projektleiter wurde ein zweiteiliger Fragebogen erstellt. Der eine Teil war das Bewerten der 24 Kriterien des Arbeitsteams. Der andere Teil des Fragebogens bestand aus den beiden Fragen:

1. Wenn Sie morgen mit einem neuen Projekt beginnen würden, welche der genannten **Kriterien** müßten aus Ihrer Sicht wesentlich verbessert sein (maximal fünf Kriterien)?

2. Welche konkreten und sofort umsetzbaren **Maßnahmen** würden Sie sich denken, um Ihr Projekt optimal - und insbesondere schnell - durchführen zu können?

Mit der ersten Frage sollte den Projektleitern die Möglichkeit gegeben werden, eine zusätzliche Gewichtung der Kriterien vorzunehmen. Die zweite Frage sollte neue Ideen für weitere Maßnahmen aufzeigen. Drei Beispiele an vorgeschlagenen weiteren Maßnahmen sind:

- Räumliche Zusammensetzung der FB- und DV-Mitarbeiter
- Projekte in kleinere Stufen zerlegen
- Konsequente Releaseplanung

3.3.4 Projektleiter-Bewertungen

Bei der **ersten Kriterienauswahl** der PL-Bewertungen wurden die Kriterien ausgesucht, bei denen der "Erfolg und Nutzen" sowie der "Erfolgszeitpunkt" mit kleiner oder gleich zwei bewertet worden sind. Dieses Verfahren ist identisch mit der Kriterienauswahl der Bewertungen von den Unternehmensberatungen. Das dazugehörige Ergebnis wird im weiteren Verlauf **"PL-Bewertung: hoher und schneller Nutzen"** bezeichnet.

Eine **zweite Kriterienauswahl** ergab sich aus der ersten Frage im PL-Fragebogen. Bei deren Auswertung haben sich nachfolgende Kriterien in der Anzahl der Nennung durchgesetzt:

- Projektziele / Qualitätsanforderungen (Kriterium 1) = 7 mal
- Softwaretechnologie (Kriterium 3) = 6 mal
- Kommunikation (Kriterium 7) = 8 mal
- Test (Kriterium 11) = 7 mal
- Bürokratie (Kriterium 19) = 5 mal
- Dokumentation (Kriterium 23) = 5 mal

Dieses Zwischenergebnis wird **"PL-Bewertung: explizite Nennung"** genannt.

Damit bei der Endauswahl der Kriterien den PL-Befragungen und den entsprechenden Istzustandsnoten mehr Bewertungsgewicht zukommt, wurde für eine zusätzliche Kriterienauswahl die Schnittmenge von den sechs Kriterien, die die Projektleiter aus ihrer Sicht für wesentlich verbesserungswürdig halten (PL-Bewertung: explizite Nennung) und den schlechten Istzuständen gebildet. Unter einem schlechten

Istzustand sind hier die Kriterien zu verstehen, die schlechter als drei bewertet worden sind. Dazugehörige Kriterien sind unter dem Namen **"PL-Bewertung: schlechter Istzustand"**, aufgeführt. Das Schnittmengenergebnis zwischen "PL-Bewertung: explizite Nennung" und "PL-Bewertung: schlechter Istzustand" wurde als **"PL-Bewertung: explizite Nennung und schlechter Istzustand"** bezeichnet.

3.4 Endauswahl der Kriterien

Für die Endauswahl der Kriterien wurden die in den verschiedenen Kriterienauswahlen zuvor herausgefundenen Ergebnisse verglichen und auf Übereinstimmungen der jeweiligen Ergebniskriterien hin überprüft. Zunächst folgt eine Übersicht mit den vielfältigen Ergebnissen.

Ergebnisherkunft (Herkunftstyp)	**Kriterien**
PL-Bewertung: explizite Nennung und schlechter Istzustand	1, 7, 11, 23
PL-Bewertung: hoher und schneller Nutzen	1, 4, 5, 7, 8, 10, 11, 15, 16, 18, 19
Unternehmensberatung: Gesamtbewertung	**1, 4, 24,** 2, 7, 10, 13, 14, 15, 18
Arbeitsteam: Gesamtbewertung	**1, 2, 4, 8, 14, 18, 19, 20,** 12, 13, 16, 17

In "Arbeitsteam: Gesamtbewertung" ist das Ergebnis "Arbeitsteam: Übereinstimmungsbewertung" mit den fett gedruckten Kriterien eingeschlossen. Die fett markierten Kriterien der "Unternehmensberatung: Gesamtbewertung" repräsentieren

Kriterien, die Übereinstimmung bei allen drei Unternehmensberatungen erzielt haben.

Als nächstes wurde das **Ergebnis** der Endauswahl der Kriterien durch den Vergleich der o.a. Ergebniskriterien ermittelt.

Übereinstimmung in	**Kriterien**
allen vier Herkunftstypen	**1 = Projektziele**
drei Herkunftstypen	**4 = FB-Beteiligung** **7 = Kommunikation** **18 = Motivation**
zwei Herkunftstypen	**2, 8, 10, 11, 14, 15, 16, 19**

Dabei ergab sich das Kriterium Projektziele als "Sieger". So hatte sich das Kriterium Projektziele aus den Bewertungen des Arbeitsteams, der Projektleiter und der drei Unternehmensberatungen unter Berücksichtigung der in der Studie verwendeten spezifischen Selektionen übereinstimmend als essentielles Kriterium "herauskristallisiert". Aber auch den Kriterien FB-Beteiligung, Kommunikation und Motivation sollte besondere Aufmerksamkeit zugewendet werden. Zu diesen vier genannten Kriterien sind als erstes wirksame Maßnahmen zur Verbesserung der vorhandenen Situationen zu finden und umzusetzen. Doch mehr zu den Maßnahmen im nächsten Kapitel.

4 Ergebnisse der Untersuchung

Ergebnisse der vom CNV-Management in Auftrag gegebenen Untersuchung sind die an das Anwendungsentwicklungsmanagement (AE) vorgetragenen Vorschläge an Sofortmaßnahmen und ein aufbereiteter Maßnahmenkatalog. Mit diesen Ergebnissen wurde die eigentliche Untersuchung abgeschlossen und vom Management eine Entscheidung zu den vorgeschlagenen Maßnahmen getroffen, um die EDV-Projektlaufzeiten in der CNV zu verkürzen.

Die Vorschläge von Sofortmaßnahmen an die AE-Runde sind **aus dem gesamten Verlauf der Untersuchung** entstanden und waren in einem **Abschlußbericht** enthalten, der dem AE-Management vorgelegt wurde. Der Inhalt des Abschlußberichtes bestand in einer skizzierten Zusammenfassung der Untersuchung. So wurden u.a. die wesentlichen Ergebnisse der Expertenbefragungen (Unternehmensberater und Projektleiter), die Endauswahl der Kriterien und die in den anschließenden Abschnitten aufgeführten Sofortmaßnahmen vorgestellt. Die Sofortmaßnahmen sind den Ergebniskriterien aus der "Endauswahl der Kriterien" zugeordnet worden.

4.1 Formale Kickoff-Veranstaltung für neue Projekte

Der Vorschlag dieser ersten Sofortmaßnahme aus dem Abschlußbericht ist dem Kriterium Projektziele - "Sieger" bei der Endauswahl der Kriterien - zuzuordnen. Ein formales Projekt-Kickoff ist von allen drei Unternehmensberatungen empfohlen worden. In den nachfolgenden Einzelpunkten dieser Sofortmaßnahme sind auch die Ideen (Maßnahmenvorschläge) des Arbeitsteams enthalten:

- Teilnehmer: Kernteam des Projektes inkl. Fachbereich.
- Teambildungsworkshop zusammen mit der Abteilung Personalentwicklung.
- Workshop zu den Zielen und Risiken des Projektes sowie zu den zeitlichen Vorgaben. Dabei Diskussion mit Vertretern des Top-Managements.
- Vorstellung organisatorischer Rahmenbedingungen des Projektes, wie Aufgabenzuordnung, Teilprojekte, Schnittstellen und Verantwortungen.
- Anforderungen an den Skill verifizieren.

4.2 Verbindlichkeit in der Zusammenarbeit von DV und FB

Auch diese Sofortmaßnahme ist von allen Unternehmensberatungen vorgeschlagen worden und betrifft offensichtlich das Kriterium FB-Beteiligung, welches mit Unterstützung der "PL-Bewertung: hoher und schneller Nutzen" eine Übereinstimmung in drei Herkunftstypen erzielt hat und mit nachstehenden Einzelmaßnahmen zu verbessern ist:

- Den organisatorischen Rahmen projektspezifisch ausrichten.
- FB- und DV-Projektleiter **verpflichten** sich in einem gemeinsamen Zeit- und Ressourcenplan.
- Eindeutige Ansprechpartner und geklärte Zuständigkeiten in der Zusammenarbeit von DV und FB festschreiben.
- Räumliche Nähe der Teammitglieder - auch zeitweise - organisieren.
- "Zuhause" des Projektes durch Standardbesprechungsraum, Projektlogo, Modellbüro usw. schaffen.

Neben den o.a. sofort umsetzbaren Einzelmaßnahmen besteht das langfristige Ziel dieser Sofortmaßnahme darin, daß **alle** Projektmitarbeiter **fulltime** mit im Projekt sitzen.

4.3 Aufbereiteter Maßnahmenkatalog

Der erstellte **Maßnahmenkatalog** enthält sämtliche Maßnahmen, die im Laufe der Studie mit der Hilfe des Arbeitsteams, der Projektleiter und der drei Unternehmensberatungen erarbeitet worden sind. Mit der Befragung der internen Experten konnten ferner bestehende Probleme festgestellt werden, denen im Maßnahmenkatalog die in der Untersuchung gefundenen Maßnahmen zugeordnet worden sind. Der aufbereitete Maßnahmenkatalog wurde an ein CNV-Gremium weitergeleitet. Dieses Gremium analysiert, bewertet, gibt entsprechende Aufträge an die zuständige Stellen weiter und kontrolliert die Umsetzung der verteilten Aufräge.

5 Ausblick auf weitere Schritte

Als weiterführender Schritt der Untersuchung ist - gemäß der vom Management getroffenen Entscheidung - die vorgeschlagene Sofortmaßnahme "Kickoff-Veranstaltung für neue Projekte" für die Umsetzung in der CNV zu organisieren. Aber auch die anderen Maßnahmen sind sukzessive zu verwirklichen. Ferner sollten die Befragungen der Projektleiter regelmäßig, z.B. jährlich, wiederholt werden, um ein feedback zu den durchgeführten Verbesserungsmaßnahmen zu erhalten und um ggf. neue Verbesserungspotentiale zu ermitteln. Damit können im Laufe der Zeit Erfahrungen gesammelt werden, mit denen die einsetzenden Maßnahmen meßbar gemacht werden. Denn in der Untersuchung fand keine wesentliche Berücksichtigung der Wirtschaftlichkeit und des Verhaltens der EDV-Projektlaufzeit im Spannungsfeld der Projektabwicklung - "Magisches Dreieck" - statt, da das Management der Laufzeitverkürzung erste Prioriät beigemessen hatte.

Eines jedoch wird bei dem Ziel der Verkürzung von EDV-Projektlaufzeiten bestehen bleiben, nämlich eine multifaktorielle Problematik für die es (vorerst) kein Erfolgskonzept gibt. Deswegen ist die Forschung aufgefordert, herauszufinden, welche Maßnahmen unter welchen Bedingen besonders zu empfehlen sind.

Literatur

[Bun97] Bundschuh, Manfred: Projekterfolgs- / Mißerfolgskriterien, in: Projektmanagement-Fachmann, RKW/GPM (Hrsg.), 1997

[Jon96] Jones, Capers: Patterns of Software Systems failure and Success; Bonn, Boston, London: International Thomson Computer Press, 1996, S. 9, 11

[Sch96] Schmidt, Peter: Effektivere Softwareentwicklung - TQM als Sicherungskonzept; Client Server Computing 12/96, S. 47

Wissensbasierte Techniken zur Flexibilisierung von Workflowsystemen

Barbara Dellen, Harald Holz, Frank Maurer, Gerhard Pews

Abstract

Dieses Papier stellt einen Ansatz vor, in dem Flexibilitätsdefizite beim Einsatz herkömmlicher Workflowmanagementsysteme in Softwareprozessen gelöst werden. Zum einen werden unter Verwendung expliziter Prozeß- und Produktmodelle Planungs- und Ausführungsschritte verzahnt. Zum anderen können Pläne von Workflows während der Abwicklung geändert, die Effekte der Änderungen berechnet und betroffene Personen benachrichtigt werden. Als Grundlage dienen dabei Mechanismen zur Verwaltung von Abhängigkeiten zwischen Entscheidungen, Prozessen und Produkten.

1 Motivation

Im globalen Wettbewerb ist die Reduzierung der finanziellen und zeitlichen Kosten von Softwareprozessen oberstes Ziel. Begriffe wie *lean management* und *business process reengineering & optimization* sind in diesem Zusammenhang in den Mittelpunkt des Interesses gerückt. Als wichtiges Hilfsmittel zur Verbesserung von Softwareprozessen wurden in diesem Zusammenhang Workflowmanagement-Systeme eingeführt.

Galler, Hagemeyer und Scheer beschreiben in [GHSch95] den Lebenszyklus von Workflows. Dabei gehen die Autoren von einem *Zwei-Phasen-Ansatz* aus, in dem die Modellierung eines Workflows vor dessen Ausführung abgeschlossen sein muß. Diese Annahme ist für Softwareprozesse nicht haltbar, da hier eine größere Flexibilität unverzichtbar ist. Wainer et al. [WWV96] argumentieren:

> *„In real life, both in office and in scientific lab environments, the enactment of a workcase may deviate significantly from what was planned/modeled.“*

Galler und Scheer unterscheiden daher in [GSch95] zwischen *fallorienterten* und *adhoc* Workflows. Erstere folgen gewissen Regeln, sind aber nicht vollständig standardisierbar, wogegen letztere unstrukturierte, einmalige Prozesse beinhalten. Georgakopolous und Hornick definieren ad-hoc Workflows ähnlich:

> *„Ad-hoc workflows perform processes, where there is no set pattern for moving information among people. Ad-hoc workflows typically involve human coordination, collaboration or co-decision. Thus the ordering and coordination of tasks (activities) in an ad-hoc workflow are not automated but are instead controlled by humans. The task ordering and coordination decisions are made while the workflow is performed." [GeH95]*

Zur Unterstützung von ad-hoc Workflows werden oftmals Groupware-Ansätze vorgeschlagen. Diese haben jedoch den Nachteil, daß sie zu wenig Strukturierung und Anleitung erlauben, um Softwareentwicklungsprozesse zu steuern.

Unsere Anwendungen stammen aus dem Bereich der Bauleitplanung und des Software Engineering. Aus dieser Perspektive heraus entstand der Wunsch, die Vorteile des Workflowmanagement und der Groupware zu vereinen und deren Nachteile zu überwinden. In diesem Papier werden wir zeigen, wie wissensbasierte Techniken dazu genutzt werden können, dieses Ziel zu erreichen.

Das Papier ist wie folgt aufgebaut: Im nächsten Kapitel werden Anforderungen an eine flexible Workflowmanagement-Umgebung im Entwurfsbereich aufgezeigt und unsere Lösungsansätze vorgestellt. In Kapitel 3 wird die Architektur des von uns entwickelten Prototypen skizziert. Kapitel 4 geht auf die dem System zugrundeliegenden Techniken ein. Eine vergleichende Analyse verwandter Arbeiten erfolgt in Kapitel 5. Das letzte Kapitel faßt die Ergebnisse zusammen und zeigt zukünftigen Forschungsbedarf auf.

2 Anforderungen und Lösungsansätze

Gegenstand unserer Arbeit ist die Entwicklung von Methoden, Techniken und Softwaresystemen für die Planung und Ausführung großer Entwurfsprozesse.

Bei näherer Betrachtung unserer Anwendungsdomänen lassen sich folgende Anforderungen an ein Softwaremanagement-Werkzeug erkennen:

- Das Werkzeug muß die explizite Modellierung von Workflows erlauben. Der resultierende Plan muß *automatisch* auf ein operationales Workflowmodell abgebildet werden, da eine explizite Implementierung (wie im zweiphasigen Workflow-Ansatz) zu zeit- und kostenintensiv wäre.

- Die an einem Projekt beteiligten Menschen arbeiten an unterschiedlichen Orten und zu unterschiedlichen Zeiten an einem gemeinsamen Ziel. Die Identifikation und Koordination der einzelnen Aktivitäten ist zeit- und kostenintensiv und erfordert eine detaillierte Projektplanung.
- Bei lang andauernden Entwicklungsprozessen ist es nicht möglich, einen detaillierten Projektplan gemäß dem zweiphasigen Workflow-Ansatz vorab zu erstellen. Viele Entscheidungen können erst unter Kenntnis der situativen Bedingungen (wie z. B. Verfügbarkeit von Agenten) oder aufbauend auf Ergebnissen vorangegangener Prozesse gefällt werden.
- Entwurfsprozesse dauern oft Monate bis Jahre und unterliegen daher zwangsläufig sich ändernden Anforderungen und Rahmenbedingungen; desweiteren können sich Entscheidungen als fehlerhaft herausstellen. Das Werkzeug sollte daher ein Änderungsmanagement bereitstellen, das Änderungen des Workflows erlaubt, betroffene Teile des Projekts identifiziert, betroffene Mitarbeiter informiert und das Projekt möglichst automatisch wieder in einen konsistenten Zustand zurückführt.

Um die obigen Anforderungen erfüllen zu können, haben wir wissensbasierte Techniken und Workflowmanagement-Techniken integriert. Folgende Entwurfsentscheidungen wurden getroffen:

Explizite Prozeß- & Produktmodelle: Ausgehend von Knowledge- und Software Engineering-Ansätzen [BV94, BLRV95, Ver94] haben wir eine Sprache zur Beschreibung von Projektplänen entwickelt.

Bereitstellung einer Workflowmaschine: Eine Workflowmaschine erhält Projektpläne als Eingabe und koordiniert deren Abwicklung. Dazu gehört die Verwaltung des aktuellen Projektzustands und der im Projektverlauf entstehenden Produkte, die Bereitstellung von Arbeitslisten für die Benutzer und die Weiterleitung wichtiger Information an die Benutzer.

Verzahnte Planung und Ausführung: In CoMo-Kit können alternative Vorgehensweisen zum Erreichen eines Prozeßziels modelliert werden. Bei der Abwicklung entscheidet sich der Benutzer für die im aktuellen Projektzustand geeigneteste Vorgehensweise. Desweiteren kann die Menge der alternativen Vorgehensweisen jederzeit erweitert werden.

Automatische Generierung von Abhängigkeiten aus dem Projektplan: Um die Voraussetzung für ein umfassendes Änderungsmanagement zu schaffen, wurde eine generische Abhängigkeitstheorie in die Workflowmaschine integriert, mit der automatisch Abhängigkeiten aus den Projektplänen abgeleitet werden können. Durch die Kenntnis der Abhängigkeiten können von Änderungen abhängige Teile des Projektes und betroffene Mitarbeiter identifiziert werden.

Änderung von Planungsentscheidungen „on the fly“: Da die Workflowmaschine den aktuellen Projektplan als Datenmenge und nicht als Programmcode ansieht, kann jener zu einem beliebigen Zeitpunkt geändert werden. Der Einsatz wissensbasierter Techniken ermöglicht dabei eine weitestgehend automatische Aktualisierung des Projektzustandes und die Notifikation betroffener Bearbeiter.

3 Architekturskizze von CoMo-Kit

Die entwickelten Techniken wurden in einem prototypischen System umgesetzt. Im folgenden wird kurz auf dessen Architektur und technische Aspekte eingegangen.

Abbildung 1 zeigt die Komponenten des Systems graphisch. Das System besteht aus einem Modellierungswerkzeug zur Erstellung von Projektplänen (*Modeler)* und einer Workflowmaschine (*Scheduler*). Ein Projektplan enthält Wissen über auszuführende Prozesse, die Struktur und den Typ der auftretenden Produkte, Produktflußabhängigkeiten, Lösungswege und beteiligte Personen. Für jedes Projekt wird ein neuer Scheduler erzeugt, der die Abarbeitung eines Projektplans koordiniert. Scheduler und Modeler sind eng miteinander gekoppelt. Zum einen wird jede Modelländerung dem Scheduler mitgeteilt, so daß dieser seine Sicht auf den Plan aktualisieren kann; zum anderen organisiert der Scheduler Auswahl und Rückzug von Entscheidungen, die den in Ausführung befindlichen Plan verändern.

Der Scheduler übernimmt darüber hinaus die Kommunikation mit den ausführenden Agenten über Clientprozesse. Der Nachrichtenaustausch ist bidirektional. Aufgabenlisten, einzusehende Produkte und Informationen über die Objekte des Projektes werden dem Agenten durch den Scheduler zur Verfügung gestellt. Der Agent teilt dem Scheduler die Ergebnisse seiner Aktivitäten und Entscheidungen mit.

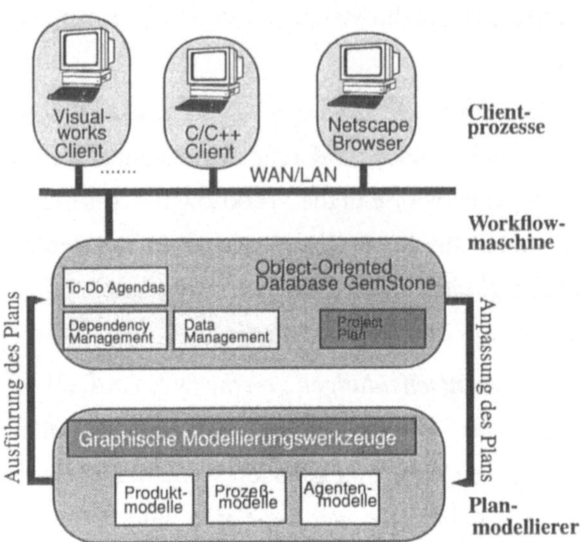

Abbildung 1: Die Architektur des Systems

4 Projektplanung und Abwicklung in CoMo-Kit

Die bereitgestellte Modellierungssprache erlaubt die Definition von Prozeßmodellen, von hierarchischen, objekt-orientierten Produktmodellen und von Ressourcenmodellen [DMMV97]. Abbildung 2 zeigt einen Ausschnitt aus der Prozeßhierarchie (Fenster links oben) und dem Produktfluß (überdecktes Fenster) eines nach dem IEEE Standard [IEEE1074] entwickelten Projektplans für die Entwicklung von Softwareprodukten. Die Prozeßhierarchie enthält die Beschreibung einzelner Prozesse („Was muß getan werden?") sowie für jeden Prozeß eine Menge alternativer *Methoden* („Wie geht man vor?"). Desweiteren können Prozessen und Methoden *Vor-* und *Nachbedingungen* sowie *Qualifikationsanforderungen* an bearbeitende Agenten (nicht abgebildet) zugeordnet werden. Eine *komplexe Methode* zerlegt einen Prozeß in Teilprozesse. Sie definiert darüber hinaus den Produktfluß zwischen den Teilprozessen. *Formale Parameter* dienen dabei als Platzhalter für erzeugte und konsumierte Produkte. Durch die Ausführung einer *atomaren Methode* wird ein Produkt erstellt.

Ein Projektplan ist eine generische Struktur, welche im Projektverlauf durch Entscheidungen der Agenten konkretisiert wird. Diese Entscheidungen werden im Scheduler verwaltet. Wir unterscheiden zwischen *Entscheidungen zur Planverfeinerung*, *Abwicklungs-* und *Delegierungsentscheidungen*. Eine Entscheidung zur Planverfei-

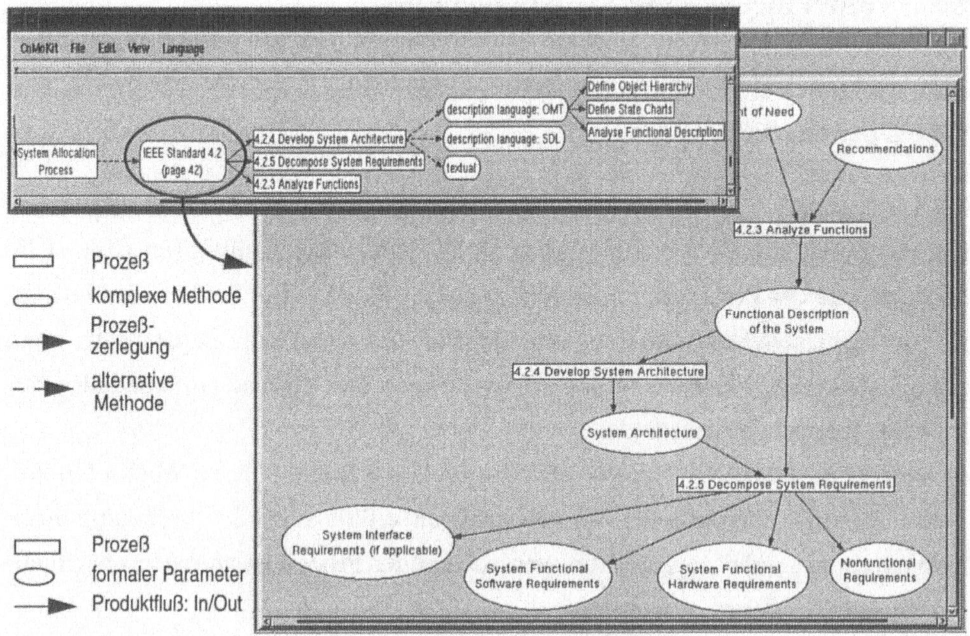

Abbildung 2: Prozeßmodellierung

nerung repräsentiert die Auswahl einer komplexen Methode. Eine Abwicklungsentscheidung wird bei der Anwendung einer atomaren Methode getroffen. Mit einer Abwicklungsentscheidung werden die Ausgabeparametern eines Prozesses mit Produkten belegt. Durch Delegierungsentscheidungen werden den Prozessen verantwortliche Bearbeiter aus dem Pool der qualifizierten Agenten zugeordnet.

Die Aktivitäten der Benutzer ändern den Zustand des Projektes. Um Entscheidungen verfolgen und nach Änderungen in einen alten Projektzustand zurückkehren zu können, muß die Workflowmaschine den Ablauf des Projektes protokollieren. Wir haben dazu Zustandsmodelle für Prozesse, Methoden, Entscheidungen und Parameter definiert. Durch die Integration kausaler Abhängigkeiten können die globalen Effekte lokaler Zustandsänderungen berechnet und Zustandsanpassungen automatisch durchgeführt werden. Der im folgenden benutzte Begriff „Objekt" ist ein Sammelausdruck für einen Prozeß, eine Methode, einen Parameter oder eine Entscheidung.

4.1 Protokollierung des Projektzustands

Im folgenden stellen wir Auszüge aus den Zustandsmodellen dar. Ein Zustandsübergang wird durch ein Ereignis (eine Benutzeraktion oder der Zustandswechsel eines abhängigen Objekts) ausgelöst.

4.1.1 Prozesse

Ein Prozeß hat die zwei Basiszustände gültig (*valid)* und ungültig (*invalid*). *Valid* bedeutet, daß der Prozeß zur Bearbeitung freigegeben ist. Ein Prozeß wird *invalid*, wenn er nicht mehr Bestandteil des aktuellen Plans ist. Dies kann z.B. passieren, wenn die übergeordnete Methode zurückgezogen wurde. Der Zustand *valid* enthält unter anderem folgende Teilzustände:

acceptable for planning: Der Prozeß kann geplant werden. Für den Wechsel in diesen Zustand müssen zwei Voraussetzungen erfüllt sein: die Vorbedingungen konnten zu „wahr" ausgewertet werden und der Prozeß ist an einen oder mehrere Agenten delegiert worden.

accepted for planning: Dieser Zustand wird erreicht, wenn ein Agent den Prozeß zur weiteren Planung selektiert hat.

performing: Von dem planenden Agenten wurde eine Methode ausgewählt und delegiert, jedoch noch nicht angewendet.

satisfied: Eine Methode wurde erfolgreich angewendet. Dieser Zustand kann wieder verlassen werden, wenn sich die angewendete Methode als ungeeignet erweist.

unsatisfied: Eine ausgewählte Methode konnte nicht angewendet werden, z. B. da die Methode im aktuellen Kontext nicht geeignet war. In dieser Situation kann vom planenden Agenten ein Rücksprung in *accepted for planning* ausgelöst und anschließend eine andere Methode ausgewählt werden.

In Abbildung 3(a) ist der Zustandsgraph für einen Prozeß dargestellt.

4.1.2 Methoden

Eine Methode hat die zwei Basiszustände *selected* und *retracted.* Diese Zustände zeigen an, ob die Methode Teil des aktuellen Plans ist oder nicht. Es sind Zustandübergänge in beide Richtungen möglich. Der initiale Zustand einer Methode ist *retracted.* Wenn bei der Bearbeitung eines Prozesses eine Methode ausgewählt wird, wechselt

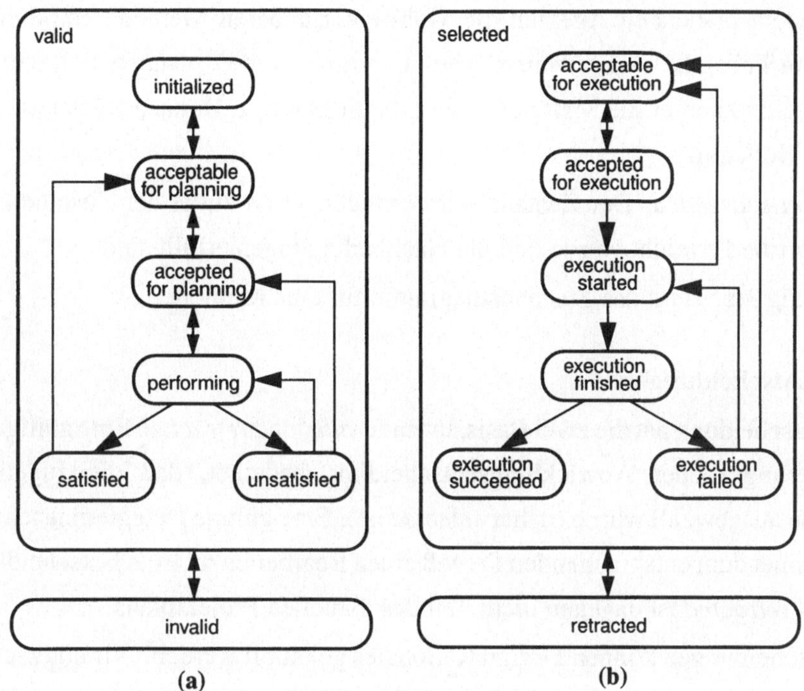

Abbildung 3: Zustandsdiagramm (a) eines Prozesses und (b) einer Methode

diese in den Zustand *selected*. Der Zustand *selected* zerfällt in verschiedene Teilzustände (Auszug):

acceptable for execution: Der planende Agent hat sich für diese Methode entschieden. Sie kann von einem ausführenden Agenten durchgeführt werden, sobald sie delegiert wurde und die Vorbedingungen zur Methodendurchführung erfüllt sind.

execution started: Der Agent hat mit der Durchführung der Methode begonnen. Die Methode bleibt solange in diesem Zustand, wie der Agent seine Arbeiten noch nicht beendet (oder abgebrochen) hat.

execution failed: Zwei Ereignisse führen in diesen Zustand. Entweder hat der Agent die Durchführung der Methode abgebrochen, oder die Nachbedingungen konnten im Zustand *execution finished* nicht zu „wahr" ausgewertet werden. Aus diesem Zustand führen benutzergesteuerte Aktionen heraus. Hatte der Agent die Durchführung der Methode abgebrochen, kann er sie in den Zustand *acceptable for execution* setzen und damit für andere Agenten zur Bearbeitung freigeben oder dem planenden Agenten die Wahl einer anderen Methode empfehlen. In diesem Fall gibt er die weitere Verantwortung an den planenden Agenten ab. Alternativ kann er die Methode erneut durchführen, z. B. nach Rücksprache mit einer Kollegin.

execution succeeded: Der Zustand wird erreicht, wenn zuvor der Zustand *execution finished* erreicht wurde und die Nachbedingungen erfüllt sind.

Abbildung 3(b) zeigt das Zustandsdiagramm für eine Methode.

4.1.3 Entscheidungen

Eine Entscheidung hat die zwei Basiszustände *valid* und *retracted.* Eine gültige Planverfeinerungs- oder Abwicklungsentscheidung bedeutet, daß die zugeordnete Methode ausgewählt wurde (daher *selected* ist). Eine gültige Delegierungsentscheidung ordnet dem entsprechenden Prozeß einen Bearbeiter zu. Eine Entscheidung im Zustand *retracted* ist dagegen nicht Teil des aktuellen Projektplans.

An Entscheidungen können Design Rationales geknüpft werden. Wir unterscheiden zwischen *harten* und *weichen* Rationales. Harte Rationales für oder gegen eine Entscheidung zwingen diese in den Zustand *valid* oder *retracted.* Über weiche Rationales für oder gegen eine Entscheidung wird ein Agent lediglich informiert. Sie bewirken keine Zustandsänderungen. Im folgenden werden die wichtigsten Zustände einer Entscheidung erläutert:

hard rationales for validity not existing: Die Entscheidung ist gültig, es existieren jedoch keine harten Rationales für die Entscheidung. Die Entscheidung kann daher ungültig gemacht werden.

hard rationales for retraction not existing: Die Entscheidung ist ungültig. Da keine harten Rationales gegen die Entscheidung sprechen, kann diese gültig werden.

hard rationales for validity existing: Die Entscheidung ist gültig und muß dies auch bleiben, da harte Rationales für diese existieren.

hard rationales for retraction existing: Die Entscheidung ist ungültig und kann nicht in den aktuellen Plan integriert werden, da harte Rationales gegen sie sprechen.

Der Scheduler stellt dabei sicher, daß nicht gleichzeitig harte Rationales für und gegen eine Entscheidung sprechen.

4.2 Ableitung und Verwaltung von Abhängigkeiten

Zwei Arten von Ereignissen können Zustandswechsel in einem Objekt auslösen: Aktivitäten der Benutzer (z.B. das Treffen einer Entscheidung) und Zustandswechsel von Objekten, die in kausalem Zusammenhang zu dem betrachteten Objekt stehen. Kausale Abhängigkeiten werden automatisch aus dem Prozeßmodell abgeleitet und im Scheduler verwaltet. Da die abgeleiteten Abhängigkeiten über- oder unterspezifiziert sein können, kann der Benutzer die Menge der Abhängigkeiten erweitern oder einschränken. Im folgenden wird ein grober Überblick über die verwalteten Abhängigkeiten gegeben. In [DKM96] beschreiben wir die Abhängigkeiten durch eine Menge von Implikationen, die von der Workflowmaschine während der Abwicklung generiert und ausgewertet werden.

Abhängigkeiten im Produktfluß: Bei der Bearbeitung eines Prozesses werden Produkte konsumiert und produziert. Bei der Operationalisierung der Pläne werden zwei Typen von Abhängigkeiten etabliert: (i) Abhängigkeiten zwischen den von einem Prozeß konsumierten Produkten und Entscheidungen zur Methodenauswahl und (ii) Abhängigkeiten zwischen Abwicklungsentscheidungen und erzeugten Produkten.

Beim ersten Abhängigkeitstyp gehen wir davon aus, daß die zu treffende Entscheidung vom aktuellen Kontext abhängt. Dies bedeutet, daß die Entscheidung für die Wahl einer Methode von den konsumierten Produkten abhängig ist. Ändern sich die Eingabeprodukte, muß die Entscheidung überdacht und gegebenenfalls revidiert werden.

Durch Abwicklungsentscheidungen werden die Ausgabeparameter der bearbeiteten Prozesse mit Produkten belegt. Die Gültigkeit dieser Parameterbelegung ist daher abhängig von der Gültigkeit der atomaren Methode. Mit der Revidierung der Entscheidung wird diese Belegung rückgängig gemacht. Das Produkt und die etablierte Abhängigkeit werden jedoch nicht gelöscht. Selektiert der Benutzer die Methode erneut, werden die alten Belegungen automatisch wieder gültig.

Abhängigkeiten zwischen einem Prozeß und seinen Teilprozessen: Mit der Entscheidung für eine komplexe Methode (Entscheidung zur Planverfeinerung) legt der Planer fest, welche Teilprozesse auszuführen sind. Die resultierenden Teilprozesse bleiben nur so lange gütig, wie die komplexe Methode Teil des Plans ist.

Delegierung: Bei der Delegierung wird die Bindung eines Agenten an einen Prozeß von der Delegierungsentscheidung abhängig gemacht. Beim Rückzug der Entscheidung wird diese Bindung wieder aufgelöst. Dadurch schafft man die Voraussetzung, Agentenbindungen bei geänderten Rahmenbedingungen (z. B. Krankheit eines Agenten) modifizieren zu können.

Produkte und Teilprodukte: Parameterhierarchien spiegeln die hierarchische Struktur von Produkten wider. Die Belegung eines Parameters setzt sich daher aus den Belegungen seiner Teile zusammen.

Benutzerdefinierte Abhängigkeiten: Die bisher aufgeführten Abhängigkeiten werden automatisch aus dem Projektplan generiert. In der Praxis zeigt es sich, daß Projektpläne unter Umständen nicht alle oder überflüssige Abhängigkeiten vorgeben. Wir haben daher Techniken entwickelt, Design Rationales während der Abwicklung definieren und durch den Scheduler verwalten lassen zu können. Für eine detaillierte Darstellung siehe [DKM96].

4.2.1 Beispiel

Untenstehend sind einige Implikationen beispielhaft für die in Abbildung 2 modellierte Methode *IEEE Standard 4.2* aufgeführt, die während der Abarbeitung generiert werden.

selected(*IEEE Standard 4.2*) ⇒
valid(*4.2.4 Develop System Architecture*) ∧
valid(*4.2.5 Decompose System Requirements*) ∧
valid(*4.2.3 Analyse Functions*) (Eq 1)

selected(*Atomic: 4.2.3 Analyse Functions*) ⇒
assigned(*Functional Description of the System* =<a functional description>) (Eq 2)

unassigned(*Functional Description of the System*=<a functional description>) ⇒
retracted(*Atomic: 4.2.3 Analyse Functions*) (Eq 3)

4.3 TMS-basierte Techniken zur Umsetzung der Konzepte

Aufgrund der Struktur der Projektpläne und der Anforderungen einer hohen Flexibilität der Entwicklungsumgebung haben sich Techniken der Aktionsplanung aus dem Bereich der Künstlichen Intelligenz als geeignet erwiesen. Ziel unserer Arbeiten ist es, den Benutzer in den Entscheidungsprozeß bei der Planung und Abwicklung einzubinden bzw. diesen selbst steuern zu lassen. Als Grundlage unserer Arbeiten diente daher REDUX [Pet91]. In REDUX werden Entscheidungen und Abhängigkeiten zwischen diesen explizit repräsentiert und verwaltet. REDUX wurde im Rahmen unseres Projektes und in Zusammenarbeit mit Dr. Charles Petrie[1] um Techniken zur Verwaltung von Delegierungsentscheidungen und Datenflußabhängigkeiten erweitert [DMPa95]. Die Zustandmodelle für Prozesse und Entscheidungen wurden ebenfalls ergänzt. Umgesetzt wurden die Zustandsmodelle und Abhängigkeitsrelationen durch ein Truth Maintenance System (*TMS)* von Doyle [Doy79]. Das TMS ermöglicht es, durch den Fortschritt des Projektes oder durch Entscheidungsänderungen bedingte Zustandswechsel entlang der etablierten Abhängigkeiten effizient zu propagieren.

5 Verwandte Arbeiten

Ein Ansatz, der die flexible Planung von Software-Entwicklungsprozessen unterstützt, ist GRAPPLE [HuLe88]. Pläne werden durch Instantiierung domänenunabhängiger und -abhängiger Operatoren erzeugt. Durch die Operatordefinitionen und die Integration von Truth-Maintenance-Techniken zur Verwaltung von vagem Wissen können die Gründe von Planungsentscheidungen dokumentiert und von Veränderungen betroffene Operatoren identifiziert werden. Zur Spezifikation von Umplanungsstrategien wird eine Metasprache bereitgestellt. Es werden jedoch keine Techniken zur Behandlung der Effekte von typischen Umplanungssituationen, wie sie z. B. von [NWC97] untersucht wurden, entwickelt.

Ein bezüglich dieser Problematik ähnlicher Ansatz ist der von Bandinelli, Fuggetta und Ghezzi [BFG93]. In der Entwicklungsumgebung SPADE wird eine reflexive Sprache SLANG bereitgestellt, in der sowohl Projektpläne als auch das Verhalten einer „Process Engine" spezifiziert werden. Die Entwicklung konkreter Abwicklungs- und Änderungsstrategien ist nicht intendiert und wird dem Benutzer überlassen. Infolgedessen wurden auch keine Mechanismen zur Behandlung der Effekte von Umplanungen entwickelt.

Ein regelbasierter Ansatz zur Beschreibung und Abwicklung von Software-Entwicklungsprozessen wird im Marvel-System [KFP88] vorgestellt. Regeldefinitionen und Objekttypen dürfen unter bestimmten Bedingungen während der Projektabwicklung geändert werden. Da kausale Abhängigkeiten zwischen Operationen und dem

1. Center for Design Research, Stanford University

Zustand von Produkten nicht explizit verwaltet werden, sind die Bedingungen sehr restriktiv.

Merlin [JPSW94] verwendet zur Definition und Abwicklung von Projekten ebenfalls einen regelbasierten Ansatz. Ein Projekt wird durch eine Menge von Fakten definiert, die in einer „Process Engine“ ausgewertet werden. Jeder Fakt gehört zu einer von fünf Klassen, die unterschiedliche Sachverhalte, z. B. kausale Zusammenhänge zwischen dem Zustand zweier Produkte, beschreiben. Die Prozeßmaschine berechnet den Folgezustand des Projekts nach Durchführung von Operationen. Der Schwerpunkt der Arbeiten liegt jedoch nicht in der Behandlung der Umplanungsproblematik.

Ein auf Graphgrammatiken basierender Ansatz wird in [HJKW96] vorgestellt. Im Gegensatz zu unserem Ansatz werden Entscheidungsänderungen durch Versionierung der Prozesse und Daten realisiert. Die Verfeinerung von Prozeßmodellen ist problemlos möglich. Durch das Versionierungskonzept stehen „Änderungen im kleinen“ im Vordergrund, d.h. Änderungen, die im wesentlichen Aktualisierungen entsprechen. Unser Ansatz betrachtet dagegen vorwiegend „Änderungen im großen“, d.h. strukturelle Änderungen, durch welche die bereits erstellten Produkte entweder überflüssig werden oder verworfen und neu erstellt werden müssen.

Im EPOS-Projekt [JC93] wurde eine detaillierte Klassifikation von Umplanungssituationen vorgenommen [NWC97]. Pläne werden von einem KI-Planer automatisch erstellt und nach Änderungen im Prozeßmodell angepaßt. Nicht operationalisiertes Wissen kann daher bei der Planung nicht berücksichtigt werden. Bei Änderungen bleibt die vollständige Rückführung des Projekts in einen konsistenten Zustand dem Benutzer überlassen.

6 Zusammenfassung und Ausblick

Durch eine explizite Planrepräsentation und die Verwaltung von Arbeitslisten entspricht CoMo-Kit dem Stand der Technik konventioneller Workflowmanagement-Ansätze. Darüber hinaus verzahnen wir die Planung und Ausführung auf drei Arten: (i) durch die zeitliche Trennung von Methodendefinition und -selektion, (ii) durch die Möglichkeit, im Modell neue Methoden zu einem Prozeß hinzufügen bzw. gegebene zu erweitern und (iii) durch Zuordnung von Verantwortlichkeiten zu Prozessen und Methoden erst bei der Abwicklung.

Weiterhin ermöglichen wir den Rückzug von Entscheidungen und die automatische Anpassung des gesamten Projektes an die geänderte Situation. Die automatische Aktualisierung des Projektes hat jedoch Grenzen. Das System kann nicht entscheiden, ob die Änderung eines Produktes den Rückzug abhängiger Entscheidungen notwendig macht. In diesem Fall werden die verantwortlichen Agenten lediglich über die Änderung informiert, wobei das System jedoch relevante Informationen zur Verfügung stellt.

Die Korrektur noch nicht ausgeführter Teile des Modells ist in CoMo-Kit problemlos möglich. Schwieriger ist die Anpassung bereits abgearbeiteter Teile des Projektes, da bereits erbrachte Ergebnisse möglichst erhalten bleiben sollen. Wir arbeiten derzeit an der Entwicklung konkreter Änderungsstrategien für solche Situationen.

Derzeit arbeiten wir an Techniken für einen verteilten Zugriff auf die Workflowmaschine über ein Wide Area Network (insbesondere Internet). Die Basis dazu bilden die von Sigrid Goldmann an der Stanford University [Go96] entwickelten Techniken. Daneben arbeiten wir auch an einer Verteilung der Workflowmaschine selber, via Local und Wide Area Networks. Die Idee ist, die entwickelten Techniken mit agentenorientierten Ansätzen, wie sie z. B. in [Pet93] vorgeschlagen werden, zu verknüpfen.

Langfristiges Ziel ist die Einbindung von Mechanismen zur Wiederverwendung von Projektplänen. Dazu müssen Pläne in einer „Experience Factory" [Bas93] verwaltet und effiziente Retrieval- und Adaptionstechniken integriert werden. Auf diese Weise würde eine weitere Voraussetzung für eine langfristige Qualitätsverbesserung von Geschäfts- und Entwurfsprozessen geschaffen.

7 Literatur

[BFG93] Sergio C. Bandinelli, Alfonso Fuggetta, and Carlo Ghezzi: Software process model evolution in the SPADE environment, in: *IEEE Transactions on Software Engineering*, 19(12), December 1993, 1128–1144

[Bas93] V. R. Basili: The Experience Factory and its Relationship to Other Improvement Paradigms, in: Ian Sommerville, Manfred Paul, eds., Proc. of the 4th European Software Engineering Conference, Lecture Notes in Computer Science Nr. 706, Springer Verlag, 1993, 68-83

[BV94] Breuker, W. van de Velde, eds.: CommonKADS Library for Expertise Modeling: IOS Press, 1994

[BLRV95] A. Bröckers, Ch. M. Lott, H. D. Rombach, and M. Verlage: MVP–L language report version 2. Technical Report 265/95: Department of Computer Science, University of Kaiserslautern, 1995

[DKM96] B. Dellen, K. Kohler, F. Maurer: Integrating Software Process Models and Design Rationales, in: Proc. of the Eleventh Knowledge Bases Software Engineering Conference KBSE 96: IEEE Computer Society, U. S., 84- 93

[DMPa95] B. Dellen, F. Maurer, J. Paulokat: Verwaltung von Abhängigkeiten in kooperativen, wissensbasierten Arbeitsabläufen, in: M. M. Richter & F. Maurer (Hrsg.), Expertensysteme 95: Proceedings in Artificial Intelligence, infix, Sankt Augustin, 1995.

[DMMV97] B. Dellen, F. Maurer, J. Münch, M. Verlage: Enriching Software Process Support by Knowledge-based Techniques: erscheint in International Journal on Software Engineering and Knowledge Engineering, 1997

[DMPe97] B. Dellen, F. Maurer, G. Pews: Knowledge Based Techniques to Increase the Flexibility of Workflow Management: to appear Data & Knowledge Engineering Journal, 1997

[Doy79] Doyle, J.: A Truth Maintenance System: Artificial Intelligence, 1979

[GHSch95] J. Galler, J. Hagemeyer, A.-W. Scheer: The Coordination of Interdisciplinary Teams in Workflow Projects, in: Proc. IDIMT-95, (Kubova Hut, Czech Rebublic)

[GSch95] J. Galler, A.-W Scheer, Workflow-Projekte: Vom Geschäftsprozeßmodell zur unternehmensspezifischen Workflow-Anwendung, in: A.-W Scheer, ed., IM-Information Management 10 1 (1995), 20-28

[Go96] S. Goldmann, Procura: A Project Management Model of Concurrent Planning and Design, in: Proc. WET ICE 96, (IEEE Computer Society Press, U. S., 1996). 177-183

[HJKW96] P. Heimann, G. Joeris, C. Krapp, B. Westfechtel: DYNAMITE: Dynamic Task Nets for Software Process Management, in: Proc. ICSE 96, IEEE Computer Society Press, U. S., 1996

[HuLe88] k. E. Huff, V. R. Lesser: An Plan-based Intelligent Assistant That Supports the Software Development Process, in: Proceedings of the Third Symposium on Software Development Environments, ACM Inc., 1988.

[IEEE1074] IEEE Standard for Developing Software Life Cycle Processes (IEEE Std 1074-1991): *Institute of Electrical and Electronics Engineers, Inc.*,New York/ USA, 1992

[JPSE97] G. Junkermann, B. Peuschel, W. Schäfer, S. Wolf: Merlin: Supporting Cooperation in Software Development through a knowledge based Environment, in: A. Finkelstein, J. KRamer, B. Niseibeh (editors), Software Process Modelling and Technology, Research Studies Press, UK, 1994

{KFP88] G.E. Kaiser P.H. Feiler, S.S. Popovich: Intelligent Assistance for Software Development and Maintenance , in: *IEEE Software*, May 1988

[JC93] M. Letizia Jaccheri and Reidar Conradi: Techniques for process model evolution in EPOS, *IEEE* Transactions on Software Engineering, 19(12):1145–1156, IEEE Computer Society Press, December 1993

[NWC97] M. N. Nguyen, A. L. Wang, R. Conradi. *Total Process Model Evolution in EPOS*: erscheint in Proc. of the Twelfth International Conference on Systems Engineering (ICSE'97), Coventry University, UK, September 1997

[Pet91] Ch. Petrie: Planning and Replanning with Reason Maintenance, Dissertation, University of Texas, Austin, 1991

[Pet93] Ch. Petrie: The Redux Server, Proc. ICICIS, Rotterdam, 1993

[Ver94] M.Verlage: Multi–view modeling of software processes, in: B. C. Warboys, ed., Proc. Third European Workshop on Software Process Technology: Springer–Verlag, 1994, 123–127.

[WWV96] Wainer, M. Weske, G. Vossen , C. M. Bauzer Medeiros: Scientific workflow systems, in: Proc. NSF Workshop on Workflow and Process Automation

Prozeßautomation im kleinen Software-Unternehmen

Paul Grünbacher, Susanne Hofer

Abstract

Dieser Artikel diskutiert Probleme kleiner Software-Betriebe und deren spezielle Anforderungen bei der Automation von Entwicklungsprozessen. Zunächst erfolgt eine Darstellung von Struktur und Hauptproblemen der österreichischen Software-Industrie. Im Anschluß daran werden Ebenen der Technologieunterstützung bei der Prozeßautomation behandelt. Schließlich erörtern wir die Einführung von Prozeßautomation und diskutieren dabei die speziellen Bedürfnisse kleiner Software-Betriebe.

1 Einleitung

Das Verständnis und die kontinuierliche Verbesserung von Entwicklungsprozessen in der Softwareentwicklung werden als Voraussetzungen zur Erhöhung der Produktqualität und zur Steigerung der Effizienz in der Entwicklung angesehen [Hum89]. Qualität soll bereits in Produkte "hineinentwickelt" und nicht erst im nachhinein "hineingetestet" werden. Dabei spielen die Definition und rechnergestützte Durchführung von Prozessen eine wichtige Rolle.

Prozeßverbesserungs- und Qualitätsmanagementprojekte können auf eine Vielzahl von Methoden, Werkzeugen und auch internationale Standards aufbauen. Viele dieser Ansätze sind jedoch aufgrund ihres Umfangs und ihrer Ausrichtung vorwiegend für mittlere und größere Unternehmen geeignet. Allerdings ist die Software-Industrie in vielen Ländern kleinbetrieblich strukturiert. In Österreich etwa beschäftigen vier von fünf Software-Betrieben weniger als zehn Mitarbeiter [Hof96]. Effektive Technologieunterstützung bei der Automation von Qualitätsmanagement- und Entwicklungsprozessen muß daher die besonderen Anforderungen von Kleinbetrieben berücksichtigen.

2 Software-Industrie in Österreich

Die österreichische *Gesamtwirtschaft* wird von Klein- und Mittelbetrieben beherrscht. Die meisten Unternehmen verfügen über einen Mitarbeiterstand von 10 bis 250 Arbeitnehmern, der Anteil der Kleinstunternehmen (0 bis 9 Mitarbeiter) ist aber im Gegensatz zu anderen europäischen Staaten eher gering [Jan96]. Laut Daten des Bundesministeriums für wirtschaftliche Angelegenheiten beschäftigen nur 0,2 Prozent der Betriebe mehr als 500 Mitarbeiter.

2.1 Struktur

Die Struktur der österreichischen Software-Industrie ist aus **Tabelle 1** ersichtlich. Interessant ist, daß der Anteil der Kleinstunternehmen im Software-Bereich signifikant höher ist als in anderen Wirtschaftsbereichen. EDV-Unternehmen beschäftigen zu über 80 Prozent weniger als 10 Mitarbeiter und liegen damit deutlich unter dem Durchschnitt der Gesamtindustrie von 12 Mitarbeitern [QoQ95].

Typ	*Beschäftigte*	*Anteil Betriebe*	*Anteil Beschäftigte*	*Bruttoproduktionswert*	*Nettoproduktionswert*
Kleinstbetrieb	0–9	84,4	24,5	18,0	16,4
Kleinbetrieb	10–49	12,4	22,7	22,8	21,7
Mittelbetrieb	50–500	3,0	31,8	36,7	36,9
Großbetrieb	> 500	0,2	21,0	22,3	25,0

Tabelle 1: Struktur der österreichischen Software-Industrie [QoQ95]

Beim Fachverband für *Unternehmensberatung und Datenverarbeitung* gab es mit Stand Jänner 1996 mehr als 6500 Berechtigungen im Bereich der Datenverarbeitung. Der Großteil dieser Unternehmen ist im kleinbetrieblichen Bereich (weniger als 50 Mitarbeiter) anzusiedeln, mehr als 50 Prozent der Betriebe verfügen über weniger als sechs Mitarbeiter [Fes95]. Laut einer Statistik von [Sch96] waren von den mehr als 1800 österreichischen Software-Betrieben 700 Einpersonen-Firmen, weitere 700 Betriebe hatten zwischen zwei und fünf Mitarbeiter, nur 50 Betriebe beschäftigten mehr als 20 Mitarbeiter.

1995 waren beim Verband Österreichischer Software-Industrie (VÖSI) 42 Unternehmen mit rund 8000 Mitarbeitern (57,5% Software-Entwicklung und Wartung, 12% Vertrieb, 10% Verwaltung, 7% Beratung) als Mitglieder registriert. Diese Betriebe, die vorwiegend im Bereich Anwender- und Individualsoftware tätig sind, erzielten einen Umsatz von mehr als 20 Milliarden Schilling.

Durch die verschärfte Wettbewerbssituation besteht auch bei Klein- und Mittelbetrieben immer mehr die Notwendigkeit zur Zertifizierung. Kleinbetriebe sind oft als Zulieferer für größere Unternehmen tätig und zertifizierte Unternehmen geben den Qualitätsdruck an die Zuliefer-Industrie weiter. In Österreich sind insgesamt 1000 Betriebe nach ISO 9000-zertifiziert, 50 davon können dem Informatik-Sektor zugeordnet werden [QOQ95].

2.2 Hauptprobleme

Bei einer im Rahmen von ESPITI (European Software Process Improvement Training Initiative) 1995-1996 in Österreich in 340 Unternehmen durchgeführten Befragung wurden das Erstellen der Anforderungsspezifikation, Projektmanagement, das Management der Kundenanforderungen und ein fehlendes Qualitätsmanagementsystem als häufigste Problembereiche genannt (vgl. **Tabelle 2**) [Hof96].

Wie man aus **Tabelle 2** weiters erkennen kann, gibt es in der Software-Branche einige atypische Problemfelder mit starker Streuung zwischen den Betriebsgrößen wie z.B. Dokumentation, Installation und Wartung, Konfigurationsmanagement und das Management der Kundenanforderungen. Folgende Prozeßbereiche bereiten in Klein- und Mittelbetrieben die größten Probleme.

- *Anforderungsanalyse und -management*: Kleinbetriebe haben mit der Erstellung der Anforderungsspezifikation die größten Schwierigkeiten. Bei der Verwaltung der Kundenanforderungen hingegen haben Klein- und Kleinstbetriebe deutlich weniger Probleme als Mittelbetriebe, was sowohl auf den engeren Kundenkontakt, als auch auf kleinere Projektgrößen zurückzuführen ist.

Problem	*1-10*	*11-50*	*51-500*	*Ø*
Requirements specification	66,1	83,3	66,7	68,7
Project management	40,0	55,6	77,8	60,8
Managing customer requirements	55,1	44,4	83,3	57,8
Lack of quality system	49,5	52,9	52,6	50,3
Lack of standards	49,5	35,3	47,4	47,6
System analysis and design	46,4	50,0	44,4	47,3
Documentation	50,5	22,2	36,8	44,7
Software and system testing	44,5	38,9	55,6	44,6
Configuration management	30,5	11,1	73,7	34,0
Software installation and maintenance	9,9	5,6	22,2	10,7
Program coding	4,6	5,6	0,0	4,1

Tabelle 2: Hauptprobleme der österreichischen Softwareindustrie (nach Betriebsgrößen in Prozent) [Hof96]

- *Projektmanagement*: Mehr als die Hälfte der Betriebe haben Schwierigkeit mit Projektplanung und Projektverfolgung, wobei die Situation wird mit zunehmender Betriebsgröße problematischer wird.

- *Qualitätsmanagementsystem*: Das Fehlen eines unternehmensweiten QM-Systems wird unabhängig von der Betriebsgröße von der Hälfte der Unternehmen beklagt. Vor allem Kleinstbetriebe können durch eine Verbesserung der Dokumentation, ein Bereich bei dem noch ein deutlicher Handlungsbedarf erkannt wurde, einen ersten Schritt in Richtung Qualitätsmanagement beschreiten.

- *Konfigurationsmanagement*: Kleinbetriebe liegen deutlich unter, Mittelbetriebe doppelt über dem Durchschnitt, was auch darauf hindeutet, daß diese Problematik im kleinen Unternehmen noch nicht entsprechend erkannt wird und die Produktstruktur weniger komplex ist.

- *Analyse/Entwurf:* Generell fällt auf, daß die frühen Phasen der Software-Entwicklung deutlich mehr Probleme verursachen als die späteren Phasen.

Ähnliche Studien zeigen, daß diese Erkenntnisse nicht nur für den österreichischen Raum gelten, sondern für den gesamten deutschsprachigen Raum zutreffen [BRZ95, Sch93]. Klein- und Mittelbetriebe liegen vor allem beim Einsatz von Methoden und Werkzeugen weit hinter größeren Software-Herstellern zurück. Der Entwicklungsprozeß ist noch zu wenig standardisiert. Qualitätssicherung, Projektmanagement und Konfigurationsmanagement werden nur beschränkt eingesetzt. Einerseits fehlt das Geld für die notwendige Investition in Hard- und Software, andererseits fehlen die personellen Kapazitäten für die erforderlichen Einführungs-, Betreuungs- und Wartungsaufwände. Vor allem Programmierbüros haben noch einen enormen Nachholbedarf, aber auch die anderen mittelständischen Software-Anbieter decken die einzelnen Bereiche nur unzureichend ab. Als Hauptargumente gegen den Technologieeinsatz werden meist die zuwenig ergebnisorientierte Ausrichtung, der hohe Einarbeitungsaufwand, sowie die für Kleinbetriebe beträchtlichen Investitionskosten genannt [Sch96].

3 Software-Prozeßautomation

Software-Prozeßautomation ist die Computerunterstützung für die explizite Definition und Ausführung (enactment) von Software-Prozessen [ACC94, Chr93, GaJ96, Mad91, PSE93, ECM93]. Wir diskutieren Software-Prozeßautomation im folgenden anhand verschiedener Ebenen der Automation.

3.1 Papierlösung

Die Prozeßdefinition existiert als Qualitätshandbuch auf Papier. Bei der Modellierung von Softwareprozessen werden Aktivitäten, Ergebnisse und Rollen mit ihren Beziehungen festgelegt [Hum89]. Durch die aufwendige Verteilung und die umständliche Handhabung besteht die Gefahr, daß die Prozeßdefinition zur reinen „Schrankware„ wird und nicht den tatsächlichen Arbeitsabläufen entspricht.

3.2 Elektronisches Handbuch

Ein elektronisches Handbuch versucht die Nachteile des Mediums Papier zu überwinden. Die Beschreibung der Prozesse ist für alle Prozeßbeteiligten On-Line abrufbar. Diese Lösung kann zwar leicht im Betrieb verbreitet werden und bietet auch komfortablere Navigations- und Suchfunktionen, die korrekte Umsetzung der definierten Prozesse kann allerdings ebenfalls nicht garantiert werden. Beispiele sind die Hypertext-Versionen des Schweizer Behördenstandards HERMES und des deutschen V-Modells [BrD93].

3.3 Beschränkte Werkzeugintegration

Produkte dieser Kategorie bieten zusätzlich zur Definition des Prozeßmodells die Möglichkeit, Werkzeugaufrufe an Aktivitäten und Ergebnisse eines Vorgehensmodells zu binden. Charakteristisch ist, daß nur der Aufruf der Werkzeuge realisiert wird, jedoch keine weitergehende Integration auf Datenebene erfolgt. Oft wird auch das projektspezifische Tailoring des Vorgehensmodells unterstützt. Beispiele für solche Werkzeuge sind WinGBS von der Firma GTI und ProcessContinuum [Pla96]. An dieser Stelle sei darauf hingewiesen, daß auch Workflow-Management Systeme die hier beschriebenen Funktionen realisieren [ChJ96].

3.4 Software-Entwicklungsumgebungen

[ECM93] definiert eine Software-Entwicklungsumgebung als "... system which provides *automated support* of the engineering of software systems and the management of the *software process*". Software-Entwicklungsumgebungen ermöglichen die Integration von Prozessmanagement, Projektmanagement und Software-Entwicklung und sind durch Daten- und Aufrufintegration von Software-Werkzeugen gekennzeichnet [ScB93]. Sämtliche Arbeitsergebnisse werden dabei in einem Objektmanagementsystem (oft auch: Repository, Entwicklungsdatenbank) verwaltet, das in der Regel auch über Versions- und Konfigurationsmanagementfunktionen verfügt [HaL93, Ley94, PCT95, Sag90]. Die Benutzer werden vom System über die nächsten anstehenden Aufgaben informiert. Werkzeuge zur Kommunikationsunterstützung zwischen Prozeßausführenden gehören ebenfalls zum typischen Funktionsumfang [Grü95]. Ein Beispiel für eine solche Entwicklungsumgebung ist MaestroII bzw. Enabler von der Firma Softlab.

3.5 Prozeßorientierte Software-Entwicklungsumgebungen

Bei einer PSEE (Process-Centered Software Engineering Environment) dient das Prozeßmodell zur weitgehenden Kontrolle der Entwicklung, wobei ein Mechanismus zur Prozeßausführung [Dow88] im Mittelpunkt steht. Dabei finden oft Ansätze wissensbasierter Systeme Verwendung. Viele solcher Umgebungen (z.B. Marvel, EPOS) befinden sich allerdings noch im Forschungsstadium.

4 Einführung von Prozeßautomation

In Abschnitt 2 wurde gezeigt, daß die Software-Industrie in Österreich von Kleinbetrieben beherrscht wird. Produktivitätssteigerungen und qualitativ hochwertige Produkte sind auch bei Kleinbetrieben der Schlüssel zum Erfolg und können durch Prozeßverbesserung und Prozeßautomation erzielt werden. Viele der für größere Organisationen entwickelten Systeme lassen sich allerdings nur schwer auf die Anforderungen der kleinen Unternehmen übertragen, weshalb ein Weg zu einer sinnvollen und auch wirtschaftlich vertretbaren Umsetzung gefunden werden muß.

Die Einführung von Prozeßautomation ist aber nur im Rahmen eines Prozeßverbesserungsprogrammes sinnvoll, das auch die organisatorischen Rahmenbedingungen berücksichtigt. Ein Projekt zur Prozeßautomatisierung wird nur dann erfolgreich sein, wenn sowohl die Geschäftsleitung, als auch die Mitarbeiter genügend in den Entscheidungsprozeß miteingebunden werden. Bei der Zielsetzung sollte man dafür sorgen, daß man klar abgesetzte Ziele verfolgt, die in einem überschaubaren und daher erreichbaren Rahmen bleiben.

4.1 Analyse des Ist-Zustands

Bemühungen um Software-Prozeßverbesserungen setzen genaue Kenntnisse über den Stand der Praxis im Unternehmen voraus. Durch ein Assessment, d.h. durch eine objektive Erhebung und Bewertung des Ist-Zustandes, ist es möglich, die Stärken und Schwächen im Software-Entwicklungsprozeß aufzudecken. Die tatsächlich gelebten Prozesse einer Organisation sind oft nicht explizit festgehalten. Durch ein Assessment mit Beteiligung der Prozeßausführenden kann auch dieses verborgene und informelle Wissen bewußt gemacht werden.

Während ein Assessment nach BOOTSTRAP [Kuv94] oder CMM [Hum89, PWC95] geprüfte Auditoren erfordert und mit hohen Kosten verbunden ist, bietet sich speziell für Klein- und Mittelbetriebe die Möglichkeit des Self-Assessments an. Eine Möglichkeit auf kostengünstige Art ein rechnergestütztes Self-Assessment durchzuführen bietet das Werkzeug SYNQUEST [StS96]. SYNQUEST ermöglicht eine Analyse der Organisaton, indem 37 Fragen mit jeweils 9 Prozeßattributen bewertet werden und daraus ein quantitatives Stärken/Schwächen-Profil erstellt wird. Damit können Aussagen sowohl über problematische Prozeßbereiche, als auch über unzureichend abgedeckte Prozeßattribute wie z.B. Verantwortung, Dokumentation, oder Werkzeugunterstützung gemacht werden. Abbildung 1 zeigt eine Auswertung eines SYNQUEST Assessments. Die Größe der Kreise steht dabei für den Grad der Abdeckung eines Prozeßbereichs.

4.2 Auswahl von Prozeßbereichen

Nach einer Analyse der Stärken und Schwächen des Unternehmes gilt es, einen Aktionsplan zur Prozeßverbesserung zu erstellen. Bei der Auswahl und Prioritätensetzung von Prozeßbereichen bieten Reifegradmodelle eine Hilfestellung, da sie eine Reihenfolge der Wichtigkeit von Prozeßbereichen implizieren. Das Software Engineering Institute der Carnegie-Mellon-University definiert im Capability Maturity Model (CMM) fünf Reifegrade mit zugeordneten Prozeßbereichen [Hum89] (siehe Abbildung 2). Empirische Untersuchungen aus den USA zeigen, daß sich dort die Mehrzahl der Unternehmen auf den Stufen eins und zwei der CMM-Skala befindet [Pla96]. Ein Blick auf die Hauptprobleme der österreichischen Software-Industrie (vgl. Abschnitt 2) bestätigt diese Studien, deckt sich diese Liste doch weitgehend mit den Prozeßbereichen der zweiten Ebene des CMM.

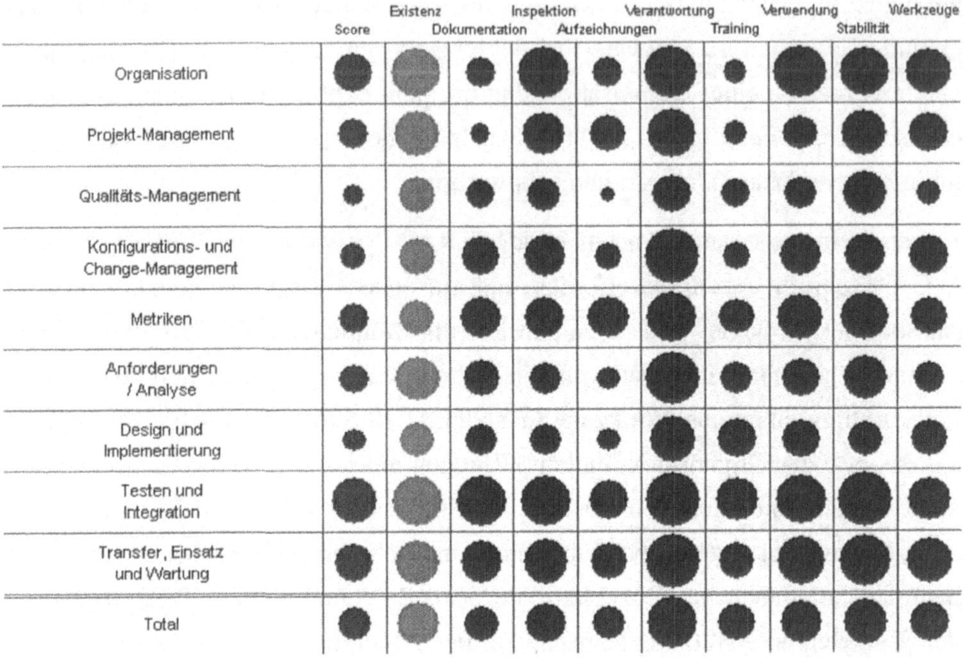

Score
Existenz
Dokumentation
Inspektion
Aufzeichnungen
Verantwortung
Training
Verwendung
Stabilität
Werkzeuge
Organisation
Projekt-Management
Qualitäts-Management
Konfigurations- und Change-Management
Metriken
Anforderungen / Analyse
Design und Implementierung
Testen und Integration
Transfer, Einsatz und Wartung
Total

Abbildung 1: SYNQUEST Auswertung Prozeßbereiche vs. Prozeßattribute

Optimizing (5)
Process change management
Technology change
Defect prevention
Managed (4)
Software quality management
Quantitative process management
Defined (3)
Peer reviews
Intergroup coordination
Software product engineering
Integrated software management
Training program
Organization process definition
Organization process focus
Repeatable (2)
Software configuration
Software quality assurance
Software subcontract management
Software project tracking and oversight
Software project planning
Requirements management
Initial (1)

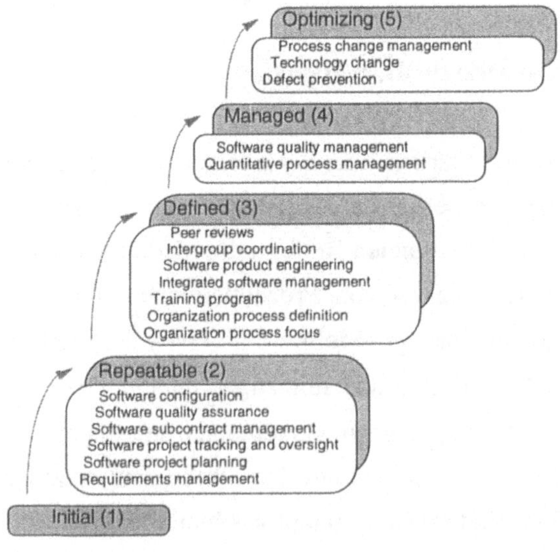

Abbildung 2: Prozeßbereiche im Capability Maturity Model [Hum89]

4.3 Voraussetzungen

Grundlegende Voraussetzung für eine erfolgreiche Realisierung der Prozeßautomation ist die Zustimmung und Unterstützung des Managements. Darüberhinaus wird nur die Automatisierung von gelebten, funktionierenden Prozessen einer Organisation zielführend sein, die folgende Voraussetzungen erfüllen:

- *Dokumentation*: Der Prozeß sollte in schriftlicher Form festgehalten sein.
- *Verantwortung:* Für den Prozeß muß die Verantwortung klar definiert sein.
- *Verwendung:* Der Prozeß sollte bereits im Unternehmen gelebt werden und sich als funktionierend beweisen.
- *Stabilität:* Erfahrungen zeigen, daß Software-Prozesse einem ständigen Wandel unterworfen sind [ACC94]. Daher sollten nur Prozesse automatisiert werden, die schon über einen längeren Zeitraum robust sind.
- *Training:* Im Unternehmen sollte ein Ausbildungskonzept das notwendige Wissen über die Prozesse vermitteln. So kann auch das Mißtrauen der Prozeßausführenden bei der Einführung neuer Technologien verringert werden.

Um Probleme während der Einführung zu minimieren, sollten zunächst nur jene Prozesse, die diese Voraussetzungen erfüllen in Form von Pilotprojekten automatisiert werden.

4.4 Besonderheiten im kleinen Software-Unternehmen

In Kleinbetrieben müssen darüberhinaus noch folgende Punkte beachtet werden:

- *Besondere Bedeutung des Tailorings*: Die Anpassung existierender Prozesse an das unternehmerische Umfeld ist unbedingt erforderlich. Umfangreiche Methoden des Qualitätsmanagements werden nur durch entsprechendes Tailoring für das kleine Unternehmen handhabbar. Im deutschen Behördenstandard V-Modell [BrD93] etwa wird das komplexe Modell in einer Tailoring-Phase an die tatsächlichen Projektbedürfnisse angepaßt.

- *Das Notwendige vor dem Nützlichen:* Bei der Auswahl der Entwicklungsumgebung bieten bestehende Kritierienkataloge [ECM93, PSE93, Her92] eine Richtschnur auch für kleine Software-Unternehmen. Die Wahl des notwendigen Automatisierungsgrades sollte immer im Zusammenhang mit den in Abschnitt 4.3 behandelten Voraussetzungen diskutiert werden.
- *Dokumentenorientiert statt aktivitätsorientiert:* Eine aktivitätenorientierte Vorgangsweise hat durch den vermehrten Einsatz von CASE-Werkzeugen und Software-Entwicklungsumgebungen an Bedeutung gewonnen. Trotz modernem Werkzeug-Einsatz bleibt die Software-Entwicklung ein dokumenten-basierter Prozeß. Die Grundlage von Spezifikation, Programmen und Dokumentation bilden immer noch Dokumente in unterschiedlichen Ausprägungen. Jeder Schritt der Software-Entwicklung wird vom Erstellen, Ändern oder überprüfen bestimmter Dokumente beherrscht und so hat sich vor allem in Betrieben, die ohne CASE-Umgebungen auskommen, eine dokumentenorientierte Software-Entwicklung durchgesetzt [WeH94]. Durch den geringeren Technologie-Einsatz ist die Arbeitsweise kleiner Softwareunternehmen stärker an den Ergebnissen als an den Aktivitäten orientiert. Bei kleinen und mittleren Projekten erscheint daher eine Instantiierung der Aktivitäten und damit verbunden eine aktivitätenorientierte Definition der Meilensteine nicht sinnvoll, die Prozeßdefinition sollte sich also stärker an den Ergebnissen z.B. in Form von Musterdokumenten orientieren [Hof96, Rüh96].

5 Zusammenfassung und Ausblick

Kleine und mittelständische Software-Betriebe verfügen über Stärken in den Bereichen technische Fachkompetenz, Spezialwissen, Flexibilität und Kundennähe. Ihre Schwächen wie unzureichendes Management, Dominanz technischen Denkens, ungenügendes Marketing, fehlende Markttransparenz und die geringe Finanzierungsbasis für Investitionen beeinflussen jedoch maßgeblich den zukünftigen geschäftlichen Erfolg.

Die Einführung von Qualitätsmanagement und Software-Prozeßautomation im kleinen Software-Betrieb erfordert die Skalierung bestehender Methoden, Standards und Technologien, um die Akzeptanz der Mitarbeiter zu gewinnen und so durch Produktivitäts- und Qualitätssteigerung einen Wettbewerbsvorteil zu erzielen. Die höhere Transparenz, bessere Dokumentation und dadurch verbesserte Nachvollziehbarkeit der Abläufe kann vor allem in Kleinbetrieben dazu führen, daß Know-How-Verluste bei Personalwechsel leichter verkraftet werden können. Die Einarbeitungszeit neuer Mitarbeiter kann dadurch ebenfalls reduziert werden.

Da die Zahl der Software-Anbieter, die speziell für Klein- und Mittelbetriebe zugeschnittene Entwicklungswerkzeuge anbieten, noch eher gering ist, sehen wir in diesem Bereich für die nächste Zeit noch ausreichendes Entwicklungspotential.

Im Rahmen des Projekts SPIRE *(Software Process Improvement in Regions of Europe)* arbeiten wir an konkreten Projekten zur Prozeßverbesserung und –automation. Dieses ESSI Projekt (Internet http://www.cse.dcu.ie/spire) richtet sich an kleine Software-Unternehmen oder Softwarearbeitsgruppen in mittleren oder größeren Betrieben anderer Branchen und wird empirische Ergebnisse über den Reifegrad der kleinen Software-Betriebe, sowie die Auswirkung von Technologieeinführung liefern. Neben Österreich nehmen daran auch Irland (Nordirland und Republik Irland), Schweden und Italien teil.

Literatur

[ACC94] B. Allgood, A. Clough, G. Cunha, J. VanBuren, S. Godfrey: Process Technologies Method and Tool Report, Software Technology Support Center (STSC), Ogden ALC/TISE, Hill AFB, Utah 84056, Bd. 1, 1994

[BRZ95] L. Bittner, C. Ricke, K. Zerr: Marketing und Vertrieb von DV-Freiberuflern und DV-Kleinunternehmen, NAA InfoKom GmbH, 1995

[BrD93] A. Bröhl, W. Dröschel (Hrsg.): Das V-Modell–Der Standard für die Softwareentwicklung mit Praxisleitfaden, Oldenbourg, 1993, ISBN 3–486–22207–4

[Chr93] A. Christie: Process-Centered Development Environments–An Exploration of Issues, Technischer Bericht, Software Engineering Institute, 1993

[ChJ96] G. Chroust, W. Jacak: Software Processes, Workflow and Work Cell Design, Proc. EUROCAST 95, Innsbruck, Springer Lecture Notes on Computer Science, 1996

[Dow88] M. Dowson (Hrsg.): Representing and Enacting the Software Process, Proc. 4th Int'l Software Process Workshop, IEEE, 1988

[ECM93] Reference Model for Frameworks of Software Engineering Environments, Technischer Bericht, European Computer Manufacturers Association, 1993

[Fes95] C. Festa: Statistik der Kammer- und Fachgruppenmitglieder 1994, Wirtschaftskammer Österreich, Abteilung für Statistik, 1995

[GaJ96] P. Garg, M. Jazayeri, (Hrsg.): Process-Centered Software Engineering Environments, IEEE Computer Society Press, 1996

[Grü95] P. Grünbacher: Software-Entwicklung als kooperativer Prozeß–Management von Interaktionsprozessen in Software-Entwicklungsumgebungen, Dissertation, Johannes Kepler Universität Linz, Abteilung Systemtechnik & Automation, 1995

[HaL93] H. Habermann, F. Leymann: Repository–Eine Einführung, Oldenbourg, 1993

[Her92] G. Herzwurm: Kriterienkatalog zur Unterstützung der Auswahl von CASE-Tools, Universität zu Köln, Wirtschaftsinformatik, Version 1.0, 1992

[Hof96] S. Hofer: Software-Entwicklung in Klein- und Mittelbetrieben: Situation, Probleme, Möglichkeiten, Dissertation, Universität Linz, Abteilung Systemtechnik und Automation, 1996

[Hum89] W. Humphrey: Managing the Software Process, Addison Wesley Reading Mass., 1989

[Jan96] H. Janik: Gemeinschaftsinitiative KMU: Operationelles Programm Österreich, In Global Village 96, Workshop Telematik für Klein- und Mittelbetriebe, 19.2.1996, Wien, OÖ Datenhighway Entwicklungs-GmbH, 1996

[Kuv94] P. Kuvaja, J. Similä, L. Krzanik, A. Bicego, S. Saukkonen, G. Koch: Software Process Assessment & Improvement–The BOOTSTRAP Approach, Blackwell Business, 1994

[Ley94] F. Leymann: Towards the STEP neutral Repository, Computer Stadards and Interfaces, 16:4:299–319, 1994

[Mad91] N. Madhavji: The process cycle, Software Engineering, pp 234–242, 1991

[PWC95] M. Paulk, C. Weber, B. Curtis, M. Chrissis: The Capability Maturity Model–Guidelines for Improving the Software Process, Addison Wesley, 1995

[PCT95] Portable Common Tool Environment (PCTE)–Standard ECMA-149, Technischer Bericht, European Computer Manufacturers Association, 1995

[Pla96] Platinum Process Continuum–Process Management, Project Management, and the Capability Maturity Model, 1815 South Meyers Road, Oakbrook Terrace, Illinois 60181, 1996, http://www.platinum.com

[PSE93] Reference Model for Project Support Environments, Technischer Bericht, NGCR Project Support Environment Standards Working Group, National Institute of Standards and Technology, Software Engineering Institute, 1993

[QOQ95] Qualität ohne Quälerei–Handouts zum ESPITI-Seminar, Forschungszentrum Seibersdorf, Wien 1995

[Rüh96] J. Rühling: Das V-Modell in der Projektpraxis, IX 6/1996:88-93

[Sag90] J. Sagawa: Repository Manager Technology, IBM System Journal, 29:2:209–227, 1990

[ScB93] D. Schefström, G. v.d.Broek (Hrsg): Tool Integration–Environments and Frameworks, volume 1, John Wiley & Sons, 1993

[Sch93] A. Schulz: Repräsentivbefragung zum Stand der Software-Entwicklung in den deutschsprachigen Ländern, Johannes Kepler Universität Linz, GES mbH Allensbach, 1993

[Sch96] E. Schoitsch: ESPITI Regional Organization Austria: User Survey Report, Forschungszentrum Seibersdorf 1996

[StS96] C. Steinmann: H. Stienen, SynQuest–Tool Support for Software Self-Assessment, Software Process -Improvement and Practice, 2:5–12, 1996

[WeH94] J. Welsh, J. Han: Software Documents: Concepts and Tools, Software Concepts and Tools, 15(1):12-25, 1994

Vergleich von Ansätzen zur Entwicklung von Workflow-Anwendungen

Roland Holten, Rüdiger Striemer, Mathias Weske

Abstract

Workflow-Management-Systeme werden entscheidende Impulse für die Gestaltung von Informationssystemen der nächsten Generation geben. Ähnlich wie Datenbanksysteme heute werden Workflow-Systeme zukünftig als Basistechnologie in komplexen Informationssystemen verwendet werden. Ansätze zur Entwicklung von Workflow-basierten Anwendungen sind somit von großem Interesse. Ausgehend von einem Vorgehens-Meta-Modell stellt dieser Beitrag zunächst wichtige in der Literatur vorgeschlagene Vorgehensmodelle zur Entwicklung von Workflow-Anwendungen einheitlich und klassifizierend dar. Darauf aufbauend erfolgt eine vergleichende Einordnung der verschiedenen Ansätze anhand von Kriterien, die basierend auf dem Meta-Modell entwickelt werden.

1 Einleitung

Die Entwicklung von Informationssystemen erfordert komplexe Entwicklungsprozesse. Ausgehend von abstrakten, betriebswirtschaftlichen Anforderungen werden in iterativen Prozessen konkrete technische Hard- und Softwarelösungen erarbeitet. Basierend auf den Erkenntnissen der Ingenieurwissenschaften wurden zur Strukturierung der Entwicklungsprozesse im Rahmen des Software Engineering zahlreiche Phasenmodelle vorgeschlagen (vgl. z.B. [Som92]). Mit dem Aufkommen neuer, zukunftsweisender Technologien stellt sich die Frage nach geeigneten Entwicklungsprozessen. Workflow-Management als zentrale neue Technologie zielt auf eine kontrollierte und automatisierte Ausführung von Geschäftsprozessen; eine Vielzahl kommerzieller Produkte [LeA94, IBM96, SNI94, Sla96] und universitärer Prototypen [Jab95, DGS95, VoW97] sind derzeit verfügbar.

Workflow-Management-Systeme werden zukünftig ähnlich wie Datenbanksysteme erheblichen Einfluß auf die Entwicklung von Informationssystemen haben [Ley96]. Die Betrachtung von Ansätzen zur Entwicklung Workflow-basierter Anwendungssysteme ist daher von vorrangigem Interesse.

In der Literatur sind eine Reihe von Vorgehensweisen zur Entwicklung von Workflow-Anwendungen beschrieben worden [Jab95, SNZ95, DGS95]. Dabei wurde untersucht, welche Aktivitäten auszuführen sind, um ausgehend von einer konkreten Problemstellung in einem Unternehmen zu einer effizienten und angemessenen Workflow-Anwendung zu gelangen. Diese Vorgehensmodelle werden oft lediglich implizit dargestellt, was eine vergleichende Analyse erschwert. Daher ist zur Explizierung der Vorgehensmodelle zunächst ein geeignetes Meta-Modell anzugeben. In Abschnitt 2 wird ein Meta-Modell für Vorgehensmodelle definiert, anhand dessen in Abschnitt 3 verschiedene Ansätze zur Entwicklung von Workflow-Anwendungen einheitlich dargestellt werden. Abschnitt 4 definiert zunächst ein im Sinne des Meta-Modells vollständiges Kriteriensystem, das für die Analyse und den Vergleich der Vorgehensmodelle herangezogen wird. Zusammenfassende Ausführung schließen den Beitrag ab.

2 Vorgehens-Meta-Modell

Zunächst werden für den Bereich der Modellierung relevante Begriffe eingeführt. Ein Modell ist ein immaterielles Abbild eines Realweltausschnitts, das für Zwecke eines Subjektes erstellt wird ([BRS95], S. 466). Liegt einem Modellsystem M2 ein Modellsystem M1 als Objektsystem zugrunde, wird also M1 als zweckorientiert ausgewählter Realweltausschnitt für M2 betrachtet, so ist M2 bezogen auf das dem Modell M1 ursprünglich zurgrundeliegende Objektsystem als Meta-Modell zu bezeichnen ([NJJ96], S. 38).

Bei der Modellierung sind demnach unterschiedliche Abstraktionsebenen zu unterscheiden. Auf der Meta-Modell-Ebene werden Komponenten von Modellen und deren Beziehungen mit ihrer Semantik sowie die Regeln zu ihrer Verwendung und

Verfeinerung festgelegt (vgl. [FeS93], S. 86). Daher ist ein Modell stets eine Instanz eines Meta-Modells. Beispielsweise ist ein konkretes Vorgehensmodell zur Entwicklung einer Workflow-Anwendung eine Instanz eines Vorgehens-Meta-Modells. Analog besteht eine Modell-Instanz-Beziehung zwischen einem konkreten Vorgehensmodell und einer Ausführung dieses Modells, d.h. der Entwicklung einer Workflow-Anwendung.

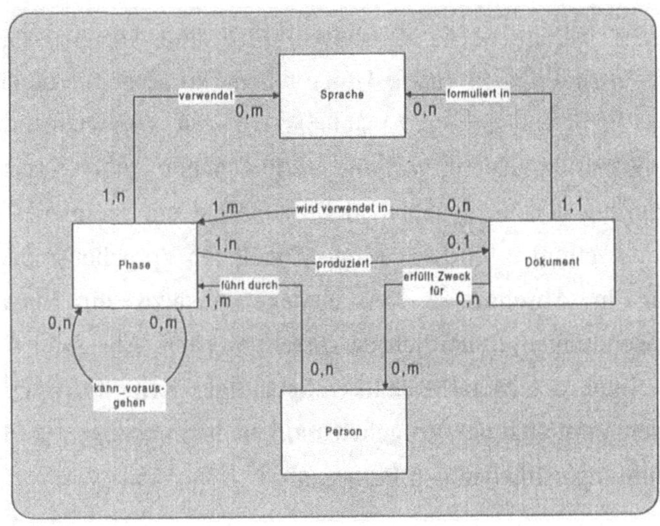

Abbildung 1: Meta-Modell zur Beschreibung von Vorgehensmodellen

Vorgehensmodelle unterscheiden eine Reihe von Phasen, in denen Dokumente erstellt werden, die in anderen Phasen verwendet werden können. Dokumente werden in Sprachen verfaßt, wobei sowohl formale Sprachen (etwa Petri-Netze oder Programmiersprachen) als auch semi-formale (graphische Sprachen wie z. B. ER-Diagramme oder Ereignisgesteuerte Prozeßketten) und nicht-formale Sprachen verwendet werden können. Tendenziell werden in frühen Phasen weniger formale Sprachen eingesetzt, in späteren Phasen werden formale Sprachen eingesetzt. Entsprechend sind eher technisch ausgerichtete Dokumente im Rahmen von konkreten Implementierungsprojekten relevant, während von technischen Details abstrahierende und betriebswirtschaftliche Problemstellungen fokussierende Dokumente im allgemeinen primär von organisatorischen Gestaltern eines Unternehmens verwendet werden. Diese Charakterisierung von Vorgehensmodellen führt zu einem Vorgehens-Meta-Modell, das in einer ER-basierten Notation in Abbildung 1 dargestellt ist. Es werden vier Entitytypen unterschieden: *Phasen*, *Sprachen*, *Dokumente* und *Personen*. Instanzen dieser Typen können in Vorgehensmodellen auf unterschiedliche Arten miteinander in Verbindung stehen. So werden die Phasen innerhalb eines Vorgehensmodells nicht isoliert voneinander in beliebiger Reihenfolge ausgeführt, sondern es besteht eine Beziehung, die eine mögliche relative Reihenfolge zwischen den Phasen festlegt; die weiteren Eigenschaften des Meta-Modells, insbesondere Beziehungen zwischen den Entitytypen, sind in Abbildung 1 dargestellt.

3 Darstellung der Ansätze

Bei der Darstellung von Ansätzen zur Entwicklung von Workflow-Anwendungen werden üblicherweise eine Reihe von Phasen unterschieden, die im folgenden kurz aufgeführt werden.

In der Phase der *Informationserhebung* werden unter Einsatz unterschiedlicher und der jeweiligen Anwendung angemessener Methoden und Techniken die Informationen der Anwendungsumgebung und der in ihr ablaufenden betrieblichen Prozesse erhoben, die für die Gestaltung betrieblicher Anwendungssysteme und insbesondere für Workflow-Anwendungen wichtig sind. In dieser Phase werden meist nichtformale Sprachen verwendet, und die erstellten Dokumente besitzen informalen Charakter.

Basierend auf den Ergebnissen der Erhebungsphase werden in der Phase der *Geschäftsprozeßmodellierung* diejenigen Daten verdichtet, die zur Identifikation, Optimierung und kontrollierten Ausführung von Geschäftsprozessen notwendig sind. In dieser Phase werden aus den zunächst noch natürlich-sprachlich beschriebenen Prozessen mittels graphischer Beschreibungssprachen semi-formal beschriebene Geschäftsprozeßmodelle erstellt und validiert. Die Geschäftsprozeßmodellierung selbst als eigenständige Phase hat die Aufgabe, ein konsolidiertes Verständnis der Ablauforganisation zu schaffen.

In der Phase der *Workflow-Modellierung* schließlich sind basierend auf semiformalen Beschreibungen von Geschäftsprozessen aus der Sicht des Workflow-Management wichtige Aspekte zu modellieren. So sind für teilautomatisiert bzw. automatisiert ausgeführte Prozesse eine Reihe von Zusatzinformationen notwendig, um das Ziel der kontrollierten Ausführung von Geschäftsprozessen zu erreichen. Es schließt sich die *Betriebsphase* an, in der Informationen anfallen, die zur weiteren Optimierung des Prozesses in der Phase der Geschäftsprozeßmodellierung verwendet werden können.

3.1 Isolierte Ansätze

Die erste Klasse von Vorgehensmodellen fokussiert insbesondere die Phase der Workflow-Modellierung. Sie wird vornehmlich von Entwicklern und Beratern durchgeführt. Aufgabe der Phase Workflow-Modellierung ist es, basierend auf den semi-formalen Beschreibungen von Geschäftsprozessen ein Workflow-Modell als Dokument zu erstellen. Workflow-Management bezeichnet dabei die kontrollierte und automatisierte Ausführung von Geschäftsprozessen. Für das Workflow-Management ist demnach insbesondere eine Abstraktion von manuell auszuführenden Teilprozessen sowie eine Konkretisierung von zu automatisierenden Teilprozessen erforderlich. Insbesondere Kontroll- und Datenflüsse müssen konkretisiert werden [LeA94, GeH95].

In der Phase der Workflow-Modellierung werden somit wesentliche implementierungstechnische Aspekte aufgenommen. Workflow-Modelle sind formale und in der Sprache von Workflow-Management-Systemen verfaßte Programme, die als Eingabe für ein Workflow-Management-System dienen und es diesem ermöglichen, zur Laufzeit die kontrollierte Ausführung der entsprechenden Workflows sicherzustellen. Kommerzielle Workflow-Management-Systeme stellen graphische Editoren bereit, mit denen Workflow-Modelle z.B. als gerichtete Graphen dargestellt werden können (z. B. [IBM96], [SNI94]). Darüber hinaus enthalten die Dokumente der Workflow-Modellierung auch rein technische Festlegungen, wie etwa konkrete Rechnernamen, Pfadnamen der Anwendungsprogramme sowie deren Parameter. Das Workflow-Modell als Dokument legt demnach die technische Integration der Hard- und Softwarekomponenten der Workflow-Anwendung fest. In Abbildung 2 sind die Phase der Workflow-Modellierung und die entsprechenden verwendeten und erzeugten Dokumente als Instanzen des Vorgehens-Meta-Modells (vgl. Abbildung 1) dargestellt. Kreise intantiieren die Meta-Entity "Dokument", Rechtecke die Meta-Entity "Phase". Durchgezogene Pfeile instantiieren die Meta-Relationships "wird verwendet in" und "produziert" zwischen "Dokument" und "Phase". Gestrichelte Pfeile instantiieren die Meta-Relationship "kann vorausgehen", also die möglichen Phasenfolgen. Der Übersichtlichkeit wegen wird auf die Darstellung der beteiligten Personen bei den Vorgehensmodell-Instanzen verzichtet. Die Betriebsphase mit ihrer Rückkopplung zur Workflow-Modellierung ist selbsterklärend und wird im Text nicht weiter erläutert.

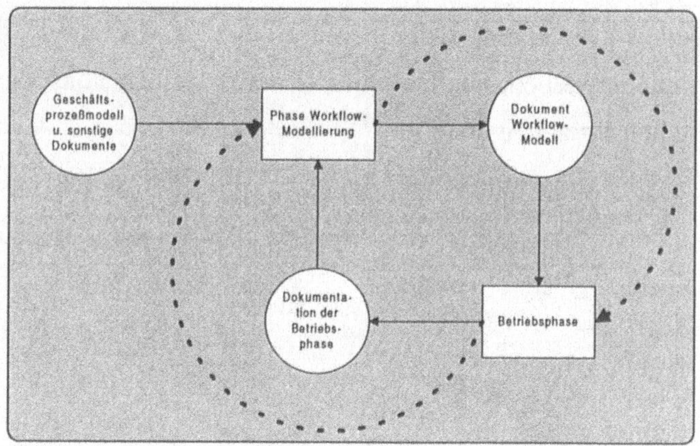

Abbildung 2: Isolierte Ansätze

3.2 Sequentielle Ansätze

Kennzeichnend für sequentielle Ansätze ist, daß die Modellierung von Workflows auf der Grundlage von Geschäftsprozeßmodellen erfolgt und diese Dokumente in Workflow-Modelle transformiert werden. Die Geschäftsprozeßmodellierung gilt als Voraussetzung für die systematische Gestaltung von prozeßorientierten Informationssystemen [SNZ95]. Geschäftsprozeßmodelle stellen in der Phase der Workflow-Modellierung Input-Dokumente dar, die gemäß den verfolgten Zwecken um workflowspezifische Informationen angereichert oder in völlig neue Dokumente transformiert werden.

Basierend auf der ARIS-Konzeption stellen [SNZ95] und [GaS94] ein Rahmenkonzept für das Management von Geschäftsprozessen vor. Die Geschäftsprozeßmodellierung und -gestaltung wird dabei als strategische Aufgabe verstanden, bei deren Ausführung eine Orientierung an den Unternehmenszielen erforderlich ist. Als Modellierungssprache für Prozeßmodelle wird in ARIS auf die erweiterten Ereignisgesteuerten Prozeßketten (eEPK) [KNS92, GaS94] zurückgegriffen. In der Phase der Prozeßkoordination, die auch als Workflow oder Workflow-Steuerung bezeichnet wird, werden die Prozeßmodelle um spezifische Informationen ergänzt und verfeinert. Rücksprünge von der Workflow-Instantiierung zur Prozeßgestaltung sind explizit vorgesehen. Sie führen stets zu erforderlichen Dokumententransformationen, da jede Phase eigene Dokumente produziert, diese jedoch inhaltlich in Beziehung zueinander stehen. Die Dokumente der einzelnen Phasen sind außerdem in unterschiedlichen Sprachen erstellt.

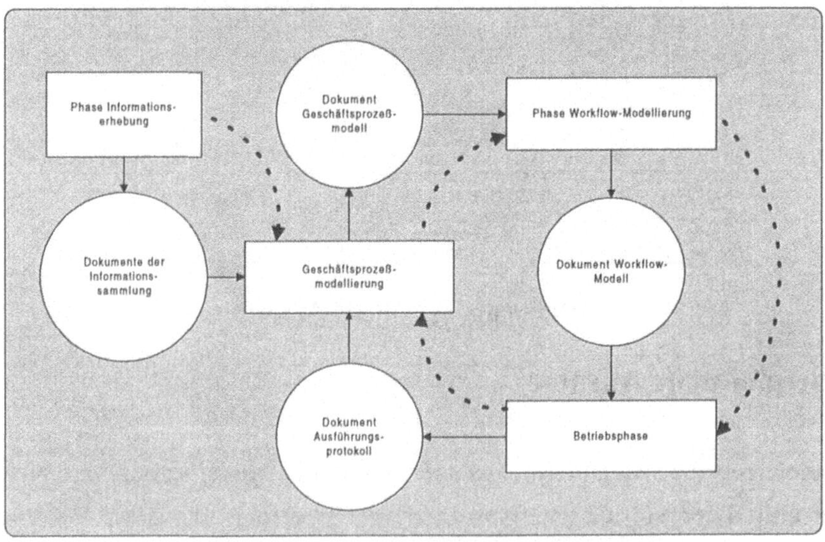

Abbildung 3: Sequentielle Ansätze

Sequentielle Ansätze der Entwicklung von Workflow-Systemen sind demnach durch drei Entwicklungsphasen charakterisiert. Die Prozeßmodellierung wird von organisatorischen Gestaltern und Informationssystementwicklern gleichermaßen ausgeführt. Darauf aufbauend führen Entwickler von speziellen Informationssystemen, nämlich Workflow-basierten Informationssystemen, die Phase der Workflow-Modellierung aus. Die Phase der Workflow-Instantiierung wird durch die Mitarbeiter des Unternehmens im täglichen Geschäft bei der Durchführung konkreter Geschäftsprozesse unter Benutzung des Workflow-basierten Informationssystems ausgeführt. Das Ausführungsprotokoll von Workflowinstanzen dient den Entwicklern der logisch vorausgehenden Phasen als Hilfsmittel bei der Prozeßanpassung. Abbildung 3 stellt die Phasenfolgen und die Dokumente der sequentiellen Ansätze in ihrem Zusammenhang dar.

3.3 Integrierte Ansätze

Integrierte Ansätze zur Entwicklung von Workflows haben zum wesentlichen Charakteristikum, daß eine Integration der Phasen des Entwicklungsprozesses über die produzierten Dokumente stattfindet. Dies bedeutet, daß alle Phasen des Vorgehensmodells sich auf eine zentrale Modellbibliothek stützen [DGS95]. Auch in den integrierten Ansätzen werden Phasen mit definierten Tätigkeiten unterschieden. So existiert zunächst eine Phase der Modellierung des Geschäftsprozesses. Das Ergebnis der Phase ist ein Geschäftsprozeßmodell, welches dann als Grundlage für eine Verfeinerung zu einem Workflow-Modell in der Workflow-Modellierungsphase dienen kann. Im Unterschied zu den sequentiellen Ansätzen wird dabei jedoch weder eine andere Sprache benutzt noch ein zweites Modell erstellt, vielmehr wird das vorhandene Modell um workflow-spezifische Informationen angereichert. Auch die Phase der Workflow-Ausführung greift auf das gleiche Modell zurück. Es erfolgt eine Interpretation des Modells, ohne daß ein Transformationsschritt durchgeführt wird. Abbildung 4 stellt das Vorgehensmodell des integrierten Ansatzes dar.

Das integrierte Vorgehensmodell erfordert die Verfügbarkeit von geeigneten Modellierungssprachen und Werkzeugen. Ein zentraler Ansatz verwendet die Sprache FUNSOFT [Gru91] und die Systeme CORMAN (wissenschaftlicher und beratender Bereich) bzw. LEU (kommerzieller Bereich) [Sla96].

4 Analytische Einordnung und Vergleich

Nach der Vorstellung der Ansätze zur Entwicklung von Workflow-Anwendungen werden die erarbeiteten Vorgehensmodelle anhand definierter Kriterien miteinander verglichen.

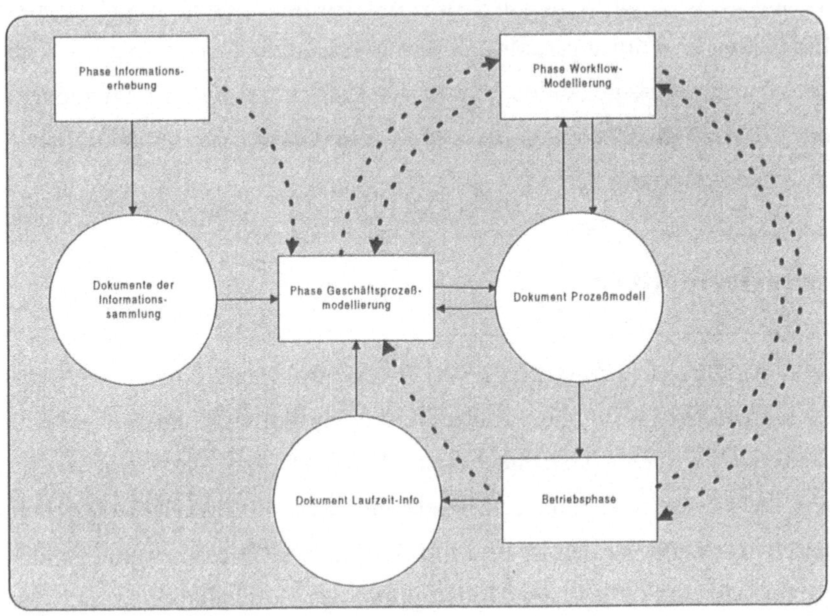

Abbildung 4: Integrierte Ansätze

4.1 Ein Kriteriensystem zur Einordnung der Vorgehensmodelle

Auf Basis des in Abschnitt 2 entwickelten Vorgehens-Meta-Modells werden nun Kriterien zur Einordnung der Vorgehensmodelle abgeleitet. Da das Vorgehens-Meta-Modell alle wesentlichen Aspekte der betrachteten Vorgehensmodelle darstellt, ist ein Kriteriensystem vollständig im Sinne des Meta-Modells, wenn alle seine Entitäten und Beziehungen durch Kriterien abgedeckt sind.

Eine wichtige Eigenschaft von Vorgehensmodellen sollte es sein, Workflow-Anwendungen mit wenig Aufwand in kurzer Zeit verändern zu können; wir bezeichnen diese Eigenschaft als *Adaptierbarkeit.* Letztlich also handelt es sich dabei um die Zeit, die für den Durchlauf eines Entwicklungszyklus benötigt wird.

Die Entities "Phase", "Sprache" und "Dokument" und ihre Beziehungen werden durch das Kriterium *Dokumentenkonsistenz* berücksichtigt. Die Forderung nach Dokumentenkonsistenz resultiert aus der Einsicht, daß die Dokumente der einzelnen Phasen eines Vorgehensmodells auch über die Funktion als Input für die jeweils nächste Phase hinaus eine Bedeutung haben. Die sich hieraus ergebende Forderung ist, daß die einzelnen Dokumente jeweils auf dem neuesten Stand sind, also bei Änderungen in einem Dokument auch andere, in Beziehung stehende Dokumente nachgezogen werden.

Die Beziehungen zwischen "Phase" und "Dokument" werden durch das Kriterium *Übergangssicherheit* abgedeckt. Als Übergangssicherheit wird die Forderung bezeichnet, daß beim Übergang zwischen einzelnen Phasen Irrtümer und Mehrdeutigkeiten möglichst ausgeschlossen werden sollten. Übergänge zwischen Phasen können eine Transformation von Dokumenten zur Folge haben. Bei einer solchen Transformation kann es u.a. zu einer mehr oder weniger ausgeprägten Umwandlung von Inhalten der Dokumente in eine andere Form kommen. Unterscheiden sich die Beschreibungsmittel signifikant, besteht die Gefahr, daß bestimmte Zusammenhänge durch die Verwendung eines anderen Beschreibungsmittels fehlgedeutet oder in einem anderen Sinnzusammenhang interpretiert werden. Solche Probleme lassen sich weitestgehend dann ausschalten, wenn eine überwiegend automatisierbare Transformation stattfindet oder keine Transformation notwendig ist.

Die "Person" steht im Mittelpunkt des Kriteriums *Zielgruppenbezogenheit.* In Projekten zur Workflow-Entwicklung arbeiten gewöhnlich eine Reihe von Personen mit unterschiedlichen Aufgaben und Erfahrungen zusammen. Dies führt zu der Anforderung, daß die erstellten Dokumente einen gewissen Grad an Zweckangemessenheit erfüllen.

Mit dem aufgeführten Kriteriensystem sind alle Komponenten des Vorgehens-Meta-Modells abgedeckt; somit ist eine vollständige Einordnung der Entwicklungsansätze anhand dieses Kriteriensystems möglich.

4.2 Vergleich der Vorgehensmodelle und Würdigung

Adaptierbarkeit

Die Beschreibungen der Ansätze in Abschnitt 3 zeigen, daß der isolierte Ansatz sicherlich die höchstmögliche Adaptierbarkeit aufweist. Durch die Tatsache, daß keine Phase der Geschäftsprozeßmodellierung vorgesehen ist (und diese somit auch bei Änderungen nicht durchzuführen ist) können Änderungen auf der Basis von Anpassungen des Workflow-Modells direkt durchgeführt werden. Der integrierte Ansatz weist einen ähnlichen Grad an Adaptierbarkeit auf. Allerdings hängt der Rücksprung im Entwicklungszyklus von der Art der durchzuführenden Änderungen ab. Sollen grundlegende Änderungen auf der Ebene der Geschäftsprozesse durchgeführt werden, ist in die Phase der Geschäftsprozeßmodellierung zu verzweigen. Im sequentiellen Ansatz dagegen ist in jedem Fall ein Rücksprung in die Phase der Geschäftsprozeßmodellierung erforderlich. Andernfalls würden evtl. Änderungen am Workflow-Modell durchgeführt, die im Geschäftsprozeßmodell nicht nachgezogen werden. Ein solches Vorgehen würde zumindest das Kriterium der Dokumentenkonsistenz verletzen, so daß jeweils ein kompletter Durchlauf des gesamten Entwicklungszyklus notwendig wird. Die Adaptierbarkeit ist für diesen Ansatz somit am geringsten.

Dokumentenkonsistenz

Als Operationalisierung der Dokumentenkonsistenz wurde die Anzahl der verschiedenen im Vorgehensmodell verwendeten Dokumententypen vorgeschlagen. Dies geschah aus der Erkenntnis, daß die Wahrung der Konsistenz verschiedener Dokumente aus Gründen der Komplexitätsbeherrschung nur bei Verwendung möglichst weniger verschiedenartiger Dokumente realistisch ist. Dies bedeutet nicht zwangsläufig, daß nur ein einziges oder wenige Dokumente, wohl aber daß wenige Dokumententypen, also Sprachen, verwendet werden. Die Nutzung derselben Sprache für unterschiedliche Dokumente kann nämlich zumindest durch automatische Anpassung von Dokumenten vorgelagerter Phasen für die Wahrung der Konsistenz eingesetzt werden. Handelt es sich dagegen um unterschiedliche Sprachen, ist eine solche Anpassung nur äußerst schwierig durchzuführen.

Für die vorstehend beschriebenen Vorgehensmodelle bedeutet dieser Zusammenhang, daß im integrierten Ansatz eine weitestgehende Dokumentenkonsistenz gegeben ist. Es existiert - abgesehen von der Phase der Informationserhebung - nur ein einziges Dokument, das demzufolge auch immer Konsistenz aufweist. Im Gegensatz dazu werden im sequentiellen Ansatz verschiedene Dokumente unterschiedlicher Sprachen ineinander überführt. Eine automatische Konsistenzwahrung ist bei Änderungen nicht möglich, da Transformationen nicht in beliebigen Richtungen möglich sind. Für den isolierten Ansatz stellt sich das Problem der Dokumentenkonsistenz nicht in der beschriebenen Form, da hier Phasen wie die Geschäftsprozeßmodellierung nicht vorgesehen sind. Bezogen auf den gesamten Entwicklungsprozeß ist die Dokumentenkonsistenz des isolierten Ansatzes aber als niedrig einzustufen.

Übergangssicherheit

Im Zusammenhang mit der Dokumentenkonsistenz ist auch die Übergangssicherheit zu sehen. Ausgangspunkt ist hier die Verhinderung von Irrtümern und Mehrdeutigkeiten bei der Transformation von Dokumenten in einer Phase des Vorgehensmodells. Betrachtet werden dabei die im Vorgehensmodell erlaubten Übergänge zwischen einzelnen Phasen. Ist es möglich, eine vollkommen automatische Transformation durchzuführen bzw. ist eine solche Transformation nicht notwendig, so ist die Übergangssicherheit gewährleistet. Dies gilt aus den im Zusammenhang mit der Dokumentenkonsistenz behandelten Gründen für den integrierten Ansatz mit Ausnahme der - in diesem Zusammenhang relevanten - Informationserhebung.

Dagegen existiert bisher kein sequentieller Ansatz, der eine eindeutige Abbildung von Geschäftsprozeßmodellen auf Workflow-Modelle erlaubt. Galler et al. stellen einen Ansatz vor, nach dem eine Ableitung von Geschäftsprozeßmodellen aus dem ARIS-Toolset in FlowMark-Modelle ermöglicht wird. Eine automatische Ableitung in die Gegenrichtung ist jedoch nicht vorgesehen. Allerdings existiert mit ContAct

ein Werkzeug, das eine automatische Benachrichtigung der Akteure in der Phase der Geschäftsprozeßmodellierung bei Änderung des Workflow-Modells übernimmt [GHS95].
Für den isolierten Ansatz gilt wiederum, daß die Übergangssicherheit weniger problematisch ist, da keine Phase der Geschäftsprozeßmodellierung vorgesehen ist. Begreift man den Ansatz als Teil eines größeren Vorhabens, in dem auch die Informationserhebung und Geschäftsprozeßmodellierung eine Rolle spielen können, ist keinerlei Übergangssicherheit gegeben.

Zielgruppenbezogenheit

Die Zielgruppenbezogenheit resultiert aus der Forderung nach zielgruppenspezifischen Beschreibungen. Sie ist dann gegeben, wenn die einzelnen, im Vorgehensmodell vorgesehenen Personengruppen durch entsprechende Dokumente in die Lage versetzt werden, die für sie relevanten Informationen gefiltert und frei von nicht relevanten Zusammenhängen zu erhalten. Für den isolierten Ansatz gilt zunächst, daß aufgrund der geringen Zielgruppenbreite (die Zielgruppe setzt sich in erster Linie aus Entwicklern und entsprechenden Beratern zusammen) die Zielgruppenbezogenheit scheinbar sehr hoch ist. Die erstellten Dokumente - im wesentlichen das Workflow-Modell - dienen den Workflow-Entwicklern hinreichend als Hilfsmittel zur Durchführung ihrer Aufgaben. Spezielle Dokumente oder Teildokumente für andere Personengruppen (etwa graphische Beschreibungen für Organisatoren oder Manager) sind nicht vorgesehen, diese Personengruppen sind im isolierten Ansatz jedoch auch nicht involviert. Daher ist die Zielgruppenbezogenheit insgesamt, unter Berücksichtigung aller für eine Entwicklung erforderlichen Phasen, als niedrig zu bezeichnen.

Läßt sich für die sequentiellen und die integrierten Ansätze hinsichtlich der Zielgruppenbreite ein ähnliches Ergebnis konstatieren (es sind die gleichen Personengruppen involviert: Mitarbeiter der Fachabteilungen, Manager, Organisatoren und Entwickler sowie Berater), zeigen sich bezüglich der Zielgruppenbezogenheit Unterschiede. Während in den sequentiellen Ansätzen eine breitere Vielfalt an zielgruppenspezifischen Dokumenten existiert (etwa Geschäftsprozeßmodelle für Mitarbeiter und Management, Workflow-Modelle für Entwickler), ist im integrierten Ansatz im wesentlichen ein einheitliches Dokument - das Prozeßmodell - für alle Personenkreise vorgesehen. Dabei existieren Konzepte zur Anpassung der Informationsmenge an die Bedürfnisse des jeweiligen Personenkreises. Zu nennen sind hier beispielsweise die Bildung von Sichten oder die Möglichkeit des "Ausblendens" bestimmter Zusammenhänge. Schlußendlich werden im integrierten Ansatz jedoch die gleichen Darstellungsmittel als Beschreibung für unterschiedliche Personenkreise verwandt, die Zielgruppenbezogenheit ist also weniger hoch als in den sequentiellen Ansätzen.

Tabelle 1 faßt den einordnenden Vergleich der Entwicklungsansätze anhand des Kriteriensystems zusammen.

Kriterium	isolierter Ansatz	sequentieller Ansatz	integrierter Ansatz
Adaptierbarkeit	hoch	niedrig	hoch
Dokumentenkonsistenz	niedrig	niedrig	hoch
Übergangssicherheit	niedrig	niedrig	hoch
Zielgruppenbezogenheit	niedrig	hoch	niedrig

Tabelle 1: Zusammenfassung der Einordnung der Entwicklungsansätze

5 Zusammenfassung und Ausblick

Die Ausführungen haben gezeigt, daß sich verschiedene Ansätze zur Entwicklung von Workflow-Anwendungen vor allem in der Anzahl der betrachteten Entwicklungsphasen und der dabei erstellten Dokumente unterscheiden. Insbesondere die Art der erzeugten und verwendeten Dokumente beeinflussen den Vergleich der Ansätze. Werden zielgruppenspezifische Dokumente erstellt, besteht die Gefahr von Inkonsistenzen bei Versionsfolgen. Wird hingegen ein stets konsistentes Zentraldokument verwandt, ist eine zielgruppenadäquate Darstellung der Zusammenhänge in den verschiedenen Entwicklungsphasen nur schwer möglich.

Bezogen auf weitere Arbeiten zur Entwicklung von Workflow-Anwendungen ist festzuhalten, daß eine vollständige Berücksichtigung aller Entwicklungsphasen erforderlich ist. Außerdem ist davon auszugehen, daß die Verwendung eines konsistenten Zentraldokumentes von großem Vorteil bei der Entwicklung ist. Allerdings sollten phasenspezifische Sichten auf dieses Dokument eine zielgruppengerechte Aufbereitung ermöglichen. Tendenziell ist also mit einem Zusammenwachsen des sequentiellen mit dem integrierten Ansatz zu rechnen. Bei der Entwicklung von Tools zur Unterstützung der Entwicklung von Workflow-Anwendungen sollten diese Tendenzen berücksichtigt werden.

Literatur

[BRS95] J. Becker, M. Rosemann, R. Schütte: Grundsätze ordnungsmäßiger Modellierung, *Wirtschaftsinformatik,* Vol. 37, No. 5, 1995, S. 435-445

[DGS95] W. Deiters, V. Gruhn, R. Striemer: Der FUNSOFT-Ansatz zum integrierten Geschäftsprozeßmanagement, *Wirtschaftsinformatik,* Vol. 37, No. 5, 1995, S. 459-466

[FeS93] O. K. Ferstl, E. J. Sinz: Grundlagen der Wirtschaftsinformatik. Band 1, Oldenbourg-Verlag, 2. Auflage, 1993

[GaS94] J. Galler, A.-W. Scheer: Workflow-Management: Die ARIS-Architektur als Basis eines multimedialen Workflow-Systems, in: A.-W. Scheer (Hrsg.): Veröffentlichungen des Instituts für Wirtschaftsinformatik, No. 108, 1994

[GeH95] D. Georgakopoulos, M. Hornick, A. Sheth: An Overview of Workflow Management: From Process Modeling to Workflow Automation Infrastructure, *Distributed and Parallel Databases*, No. 3, 1995, S. 119-153

[GHS95] J. Galler, J. Hagemeyer, A.-W. Scheer: Asynchronous Cooperation Support for Distributed Collaborative Information Modeling, in: K. Sandkuhl, H. Weber (Hrsg.): Telekooperationssysteme in dezentralen Organisationen, Tagungsband zum Workshop der GI-Fachgruppe "CSCW in Organisationen", 22.-23. Februar 1996, ISST-Bericht No. 31, 1996, S. 67-79

[Gru91] V. Gruhn: Validation and Verification of Software Process Models, Dissertation, Universität Dortmund, 1991

[IBM96] IBM FlowMark: Modeling Workflow. Version 2 Release 2, in: IBM Deutschland Entwicklung GmbH (Hrsg.), Publ. No. SH-19-8241-01, 1996

[Jab95] S. Jablonski: Workflow-Management Systeme, Thomson's Aktuelle Tutorien, Band 9, ITP, 1995

[KNS92] G. Keller, M. Nüttgens, A.-W. Scheer: Semantische Prozeßmodellierung auf der Grundlage "Ereignisgesteuerter Prozeßketten (EPK)", in: A.-W. Scheer (Hrsg.): Veröffentlichungen des Instituts für Wirtschaftsinformatik, No. 89, 1992

[Ley96] F. Leymann: The Workflow-Based/Application Paradigm, in: J. Becker, M. Rosemann (Hrsg.): Workflowmanagement - State-of-the-Art aus Sicht von Theorie und Praxis (Proceedings zum Workshop), 1996, S. 4-10

[LeA94] F. Leymann, W. Altenhuber: Managing Business Processes as an Information Resource, *IBM Systems Journal,* Vol. 33, No. 2, 1994, S. 326-348

[NJJ96] H. W. Nissen, M. A. Jeusfeld, M. Jarke, G. V. Zemanek, H. Huber: Managing Multiple Requirements Perspectives with Metamodels, *IEEE Software*, Vol. 13, No. 3, 1996, S. 37-48

[SNZ95] A.-W. Scheer, M. Nüttgens, V. Zimmermann: Rahmenkonzept für ein integriertes Geschäftsprozeßmanagement, *Wirtschaftsinformatik,* Vol. 37, No. 5, 1995, S. 426-434

[SNI94] Siemens Nixdorf: WorkParty Benutzerhandbuch Version 2.0, 1994

[Sla96] H. Slaghuis: Der direkte Übergang von BPR zu Workflow mit Leu, in: J. Becker, M. Rosemann (Hrsg.): Workflowmanagement - State-of-the-Art aus Sicht von Theorie und Praxis (Proceedings zum Workshop), 1996, S. 55-63

[Som92] I. Sommerville: Software Engineering, Addison-Wesley, 1992

[VoW97] G. Vossen, M. Weske: The WASA Approach to Workflow Management for Scientific Applications, NATO ASI Workshop, Istanbul, August 1997, erscheint in: Springer NATO ASI Series, 1997

Autorenverzeichnis

Prof. Dr. Jürgen Angele
Fachhochschule Braunschweig
Institut für Angewandte Informatik
Salzdahlumer Str. 46/48
38302 Wolfenbüttel
Tel. 05331-939-657
Fax 05331-939-602
E-Mail: angele@
informatik.fh-wolfenbuettel.de

Dr. Peter Buxmann
Institut für Wirtschaftsinformatik
J.W. Goethe-Universität Frankfurt
Mertonstraße 17
60054 Frankfurt/Main
E-Mail:
pbuxmann@wiwi.uni-frankfurt.de
http://www.wiwi.uni-
frankfurt.de/~pbuxmann

Barbara Dellen
AG Expertensysteme
Universität Kaiserslautern
Postfach 3049
67653 Kaiserslautern
E-Mail: dellen@informatik.uni-kl.de
http://wwwagr.informatik.uni-
kl.de/~comokit

Fabian Dömer
Diebold Deutschland GmbH
Frankfurter Straße 27
65760 Eschborn
E-Mail:
FabianDoemer@compuserve.com

Prof. Dr.-Ing. Peter Elzer
Technische Universität Clausthal
Institut für Prozeß- und Produktions-
leittechnik
Julius-Albert-Straße 6
38678 Clausthal-Zellerfeld
E-Mail: elzer@ipp.tu-clausthal.de

Axel Fell
Gebühreneinzugszentrale
Postfach 110363
50403 Köln
Tel. 0221-5061-2671
Fax 0221-5061-2903

Dr. Paul Grünbacher
Systems Engineering and Automation
Johannes Kepler Universität Linz
A-4040 Linz
Österreich
Tel.: (+43)-732 2468 867
Fax: (+43)-732 2468 878
E-Mail: pg@sea.uni-linz.ac.at

Helmut Häck
St. Antoniusstr. 6
51429 Bergisch Gladbach
Tel. 02204-8948

Dr. Andreas Henrich
Universität-Gesamthochschule Siegen
Fachbereich Elektrotechnik und Informatik
Praktische Informatik
Hölderlinstraße 3
57068 Siegen
E-Mail:
henrich@informatik.uni-siegen.de
http://www.informatik.uni-siegen.de/pi/user/henrich

Prof. Dr. Knut Hildebrand
Fachhochschule Ludwigshafen
Fachbereich Logistik / Wirtschaftsinformatik
Ernst-Boehe-Straße 4
67059 Ludwigshafen
Tel. 0621-5203151
Fax 0621-5203111
E-Mail: knut.hildebrand@t-online.de

Dr. Susanne Hofer
Systems Engineering and Automation
Johannes Kepler Universität Linz
A-4040 Linz
Österreich
Tel.: (+43)-732 2468 873
Fax: (+43)-732 2468 878
E-Mail: sh@sea.uni-linz.ac.at

Dipl.-Kfm. Dipl.-Inform.
Roland Holten
Westfälische Wilhelms-Universität
Institut für Wirtschaftsinformatik
Grevener Str. 91
48159 Münster
isroho@wi.uni-muenster.de

Harald Holz
AG Expertensysteme
Universität Kaiserslautern
Postfach 3049
67653 Kaiserslautern
E-Mail: holz@informatik.uni-kl.de
http://wwwagr.informatik.uni-kl.de/~comokit

Robert Hürten
Hürten & Partner Unternehmensberatung
Kirchstr. 38
64560 Riedstadt

Prof. Dr. Gerhard F. Knolmayer
Institut für Wirtschaftsinformatik der Universität Bern
Abteilung Information Engineering
Engehaldenstrasse 8
CH 3012 Bern
Tel.: +41 31 631 3809
Fax: +41 31 631 4682
E-Mail: knolmayer@ie.iwi.unibe.ch
http://www.ie.iwi.unibe.ch/staff/knolmayer

Prof. Dr. Wolfgang König
Institut für Wirtschaftsinformatik
J.W. Goethe-Universität Frankfurt
Mertonstraße 17
60054 Frankfurt/Main
E-Mail:
wkoenig@wiwi.uni-frankfurt.de
http://www.wiwi.uni-frankfurt.de/~wkoenig

Dr. Dieter Landes
Daimler-Benz AG
Forschung und Technik
Abt. Softwaregestaltung (F3K/S)
89013 Ulm
Tel. 0731-505-2869
Fax 0731-505-4210
E-Mail:
landes@dbag.ulm.DaimlerBenz.com

Axel Lukassen
Leitung IT
COLT Telecom
Eschersheimer Landstraße 10
60322 Frankfurt/Main
Tel. 069-959 58 131
E-Mail: lukassen@compuserve.com

Dr. Frank Maurer
AG Expertensysteme
Universität Kaiserslautern
Postfach 3049
67653 Kaiserslautern
E-Mail: maurer@informatik.uni-kl.de
http://wwwagr.informatik.uni-kl.de/~comokit

Günther Müller-Luschnat
FAST e.V.
Arabellastraße 17
81925 München
Tel. 089-920047-0
Fax 089-920047-18
E-Mail: gml@fast.de
http://www.fast.de/Profil/mitarbeiter.html

Gerhard Pews
AG Expertensysteme
Universität Kaiserslautern
Postfach 3049
67653 Kaiserslautern
E-Mail: pews@informatik.uni-kl.de
http://wwwagr.informatik.uni-kl.de/~comokit

Dipl.-Phys. Frank Rose
Institut für Wirtschaftsinformatik
J.W. Goethe-Universität Frankfurt
Mertonstraße 17
60054 Frankfurt/Main
E-Mail:
frose@wiwi.uni-frankfurt.de
http://www.wiwi.uni-frankfurt.de/~frose

Dr. Kurt Schneider
Daimler-Benz AG
Forschung und Technik
Abt. Softwaregestaltung (F3K/S)
89013 Ulm
Tel. 0731-505-2821
Fax 0731-505-4210
E-Mail: k.schneider@dbag.ulm.DaimlerBenz.com

Martina Schollmeyer
FAST e.V.
Arabellastraße 17
81925 München
Tel. 089-920047-0
Fax 089-920047-18
E-Mail: martina@fast.de
http://www.fast.de/Profil/mitarbeiter.html

Norbert Schulte
Organisations- und IV-Beratung
Im neuen Roth 2
47918 Tönisvorst

Dr. rer. pol. Dieter M. Spahni
Institut für Wirtschaft und Verwaltung der HWV Bern
Ostermundigenstrasse 81
CH 3006 Bern
Tel.: +41 31 332 5362
Fax : +41 31 331 5158
E-Mail: dieter.spahni@hwvbe.ch
http://www.hwvbe.ch/hwv/doz/spahni

Prof. Dr. Eberhard Stickel
Lehrstuhl für Allgemeine Betriebswirtschaftslehre, insbesondere Wirtschaftsinformatik, Finanz- und Bankwirtschaft
Europa-Universität Viadrina
Postfach 776
15207 Frankfurt/Oder
E-Mail: stickel@euv-frankfurt-o.de
http://wi.euv-frankfurt-o.de

Dipl.-Kfm. Rüdiger Striemer
Fraunhofer-Institut für Software- und Systemtechnik
Postfach 52 01 30
44207 Dortmund
ruediger.striemer@do.isst.fhg.de

Prof. Dr. Rudi Studer
Universität Karlsruhe
Institut für Angewandte Informatik und Formale Beschreibungsverfahren
Kollegiengebäude am Ehrenhof
D 76128 Karlsruhe
Tel. 0721-608-3923
Fax 0721-693717
Email studer@aifb.uni-karlsruhe.de

Dr. Mathias Weske
Westfälische Wilhelms-Universität Münster
Institut für Wirtschaftsinformatik
Grevener Str. 91
48159 Münster
weske@helios.uni-muenster.de

Werner Wirdemann
ORACLE Deutschland GmbH
Niederlassung Berlin
Wittestraße 30 N
13509 Berlin
wwirdema@de.oracle.com